Electrical Properties
of Mammalian Tissues

Electrical Properties of Mammalian Tissues

An introduction

B.J. Northover
Department of Pharmacology
Leicester Polytechnic, UK

CHAPMAN & HALL
London · Glasgow · New York · Tokyo · Melbourne · Madras

Published by Chapman & Hall 2−6 Boundary Row, London SE1 8HN

Chapman & Hall, 2−6 Boundary Row, London SE1 8HN, UK

Blackie Academic & Professional, Wester Cleddens Road, Bishopbriggs,
Glasgow, G64 2NZ, UK

Chapman & Hall, 29 West 35th Street, New York NY10001, USA

Chapman & Hall Japan, Thomson Publishing Japan, Hirakawacho Nemoto
Building, 7F, 1-7-11 Hirakawa-cho, Chiyoda-ku, Tokyo 102, Japan

Chapman & Hall Australia, Thomas Nelson Australia, 102 Dodds Street,
South Melbourne, Victoria 3205, Australia

Chapman & Hall India, R. Seshadri, 32 Second Main Road, CTT East,
Madras 600 035, India

First edition 1992

© 1992 B.J. Northover

Typeset in 11/13 Sabon by Best-Set Typesetter Ltd, Hong Kong
Printed in Great Britain by Page Bros (Norwich) Ltd, Norwich

ISBN 0 412 46050 5

A catalogue record for this book is available from the British Library

Library of Congress Cataloging-in-Publication Data

Northover, B.J., 1936−
    Electrical properties of mammalian tissues: an introduction/B.J.
Northover.
        p. cm.
    Includes bibliographical references and index.
    ISBN 0-412-46050-5
    1. Electrophysiology. I. Title.
QP341.N67   1992
599′.019127—dc20   91-40472
                                                            CIP

Contents

Preface

My encounters over the past 20 years with undergraduate and postgraduate students of various biological disciplines, both as teacher and examiner, have convinced me that a simple account of the basic electrical properties of mammalian tissues is needed. Too often, in the past, I have encountered students using, and then writing about, electrical recordings of biological events without clearly understanding basic electrical principles. It is a tribute to the designers of modern electrical recording instruments that such students can often make some progress, even under these adverse circumstances. Sooner or later, however, these students get into difficulties, most commonly over interpretations of biological events. When challenged, they have often told me that they have received little or no formal education in the physics of electricity, and that standard textbooks on the physics of electricity in the library seem divorced from biological systems that are of interest to them. As a result, such students have abandoned hope of obtaining any true understanding of biological events in electrical terms. Inadequate as that may be as an excuse, I am conscious that standard textbooks of electrophysiology tend to assume that readers are already familiar with the basic physics of electricity. A student may perhaps be excused, therefore, for thinking that nobody appreciates his or her special need. I hope that this book will meet that need.

It is only an introduction, but hopefully one that will provide a student of biology having little background knowledge of physics with a fairly painless, but rational, entry into the vast and detailed literature that now exists on various aspects of biological electricity. I am a cardiovascular pharmacologist, and this will become evident to readers in the examples chosen to illustrate basic principles of the subject. The more conventional bias of writers of books on electrophysiology has been towards neurological electricity. I hope that a change of perspective will be welcomed, not least because non-nervous, and even non-excitable, tissues of the body provide just as good a starting point. I hope also that mention of pharmacological tools that are available to probe electrical events in mammals will also provide a useful perspective for the beginner. Certainly, pharmacological agents have contributed much to the elucidation of the basic electrical properties of many mammalian tissues. At the same time it is important to recognize that the biological effects produced by such drugs in man and other animals are dependent upon interference with natural electrical phenomena in the body. Students of pharmacology, therefore, should also benefit from reading this book.

To list all the many persons who have contributed to ones own knowledge of a scientific subject is impossible, and to single out just a few of ones mentors would be invidious. It is not even possible to name all those who have helped in various practical ways with the preparation of a book like this. Nevertheless, I must mention the unstinting labours with a word processor of my wife Ann, and thank her particularly for sharing in all aspects of the long pilgrimage that is the life of any scientist.

– 1

Basic considerations

From an electrical standpoint, a mammalian cell may be envisaged as a dilute aqueous solution of certain electrolytes, separated from a surrounding aqueous medium of rather different electrolyte composition by a surface membrane known as the plasmalemma, the latter being water permeable. Proteins are mostly anionic at physiological pH and are synthesized within the cell. Because the plasmalemma is effectively impermeable to many proteins, the proteins accumulate at higher concentrations in the intracellular fluid than in the surrounding medium. Other ions, notably potassium (K^+) and chloride (Cl^-), permeate the cell membranes. For reasons to be discussed below, the K^+ and Cl^- ions also become distributed unequally across the plasmalemma. The resulting asymmetry among the permeant ions produces the varied electrical properties of all mammalian cells. The great diversity of those electrical properties should not obscure the essential uniformity of the primary cause.

Figure 1.1 depicts a typical cell interior (C_i), separated by a plasmalemma (m) from an extracellular compartment (C_o) of the same volume. The plasmalemma is permeable to K^+ and Cl^-, but not to protein anions (Pr^-) that are present as potassium salts in C_i. The permeability to K^+ and Cl^-, (and indeed to other ions) is considered to be due to ion-selective channels in the membrane (Chapter 6). If the KCl concentration [KCl] is the same originally in C_i and C_o, since the total K^+ concentration in C_i ($[K^+]_i$) has contributions from both KCl and KPr, then initially:

$$[K^+]_i > [K^+]_o \tag{1.1}$$

Diffusion of K^+ from C_i into C_o, therefore, will be faster than in the reverse direction, and so $[K^+]_i$ will decline. Nevertheless, at any particular moment the total number of anions must be the same as the total number of cations in any particular solution (compartment). Hence:

$$[K^+]_i = [Cl^-]_i + [Pr^-]_i \quad \text{and} \quad [K^+]_o = [Cl^-]_o \tag{1.2}$$

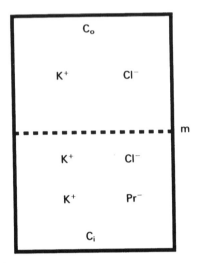

Figure 1.1 A two-compartment system, separated by a membrane (m). Aqueous fluid in the inner compartment (C_i) contains dissolved KCl and KPr, whereas that in the outer compartment (C_o) contains only dissolved KCl.

As $[K^+]_i$ falls, therefore, net Cl^- transfer from C_i into C_o will occur by diffusion. At equilibrium:

$$[Cl^-]_o > [Cl^-]_i \tag{1.3}$$

The equilibrium $[Cl^-]$ gradient created in this way represents a concentration cell. That is to say, a gradient of potential energy exists at equilibrium between C_i and C_o, because Cl^- is concentrated into C_o. The magnitude of the energy gradient will be reflected in the differing vapour pressures (P) of the solutions on either side of the plasmalemma. This is because the saturated vapour pressure of a volatile solvent, such as water, is lowered by the presence of a non-volatile solute, such as Cl^-. **Raoult's law** predicts that an inverse proportionality will exist between $[Cl^-]$ and the saturated vapour pressure of such a solution. It follows from equation 1.3, therefore, that:

$$P_i > P_o \quad \text{and} \quad \frac{P_i}{P_o} = \frac{[Cl^-]_o}{[Cl^-]_i}$$

Figure 1.1 represents a **real** mammalian cell, but it is easier to describe its real physicochemical properties if certain **hypothetical** attachments are made to it, as shown in Figure 1.2. At equilibrium, in Figure 1.2 as in Figure 1.1, solutions in C_i and C_o will have dissimilar $[Cl^-]$, with $[Cl^-]_o > [Cl^-]_i$, as predicted by equation 1.3. Via two

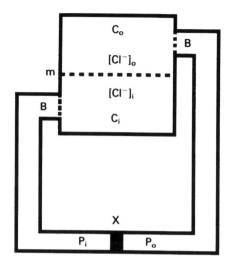

Figure 1.2 A two-compartment system, similar to that in Figure 1.1. The membrane (m) separates an aqueous solution in the inner compartment (C_i) from one in the outer compartment (C_o), such that $[Cl^-]_i < [Cl^-]_o$. Water vapour-permeable barriers (B) permit fluids in C_i and C_o to remain in equilibrium with saturated water vapour, at pressures P_i and P_o respectively. A hypothetical frictionless piston (X) separates the two water vapour-containing chambers.

hypothetical water vapour-permeable barriers (B) the two solutions are in equilibrium with two hypothetical chambers containing saturated water vapour at pressures of P_i and P_o, with the vapours separated from each other only by a frictionless piston (X). Such a piston would move rightwards spontaneously under the vapour pressure gradient. If sufficient external force was applied to the piston to move it leftwards, on the other hand, $[Cl^-]_o$ would fall and $[Cl^-]_i$ would rise, as water moved upwards through the plasmalemma and Cl^- moved in the reverse direction. The work (Q_c) necessary in order to raise P_o and to lower $[Cl^-]_o$ may be calculated as the piston is moved very slowly leftwards at constant temperature, using the equation:

$$Q_c = \int_{P_o}^{P_i} V dP \qquad (1.4)$$

where V is the volume of water vapour.

The integral sign denotes a summation of all the dP terms between the limits of $P = P_i$ and $P = P_o$. The units of Q_c would be joules per mole of Cl^- moved against the gradient ($[Cl^-]_o - [Cl^-]_i$), or against the vapour pressure gradient ($P_i - P_o$). This equation shows that the mechanical work that would need to be done in order to increase the vapour pressure gradient by a minute amount (dP) is also a function of the vapour volume (V). But, if water vapour behaves as an **ideal gas**, the **universal gas law** predicts that:

$$V = \frac{RT}{P}$$

where T is the **absolute** temperature and R is the **universal gas constant**, with a value of 8.315 joules degree^{-1} mole^{-1}.

This permits V to be substituted in equation 1.4, giving

$$Q_c = \int_{P_o}^{P_i} \frac{RTdP}{P} \quad \text{or} \quad Q_c = RT \int_{P_o}^{P_i} \frac{dP}{P} \tag{1.5}$$

Thus:

$$Q_c = RTln \frac{P_i}{P_o} \tag{1.6}$$

where *ln* denotes a **natural logarithm.**

Since vapour pressure is inversely proportional to [Cl$^-$], equation 1.6 may be rewritten as:

$$Q_c = RTln \frac{[Cl^-]_o}{[Cl^-]_i} \tag{1.7}$$

Since the solute involved is ionic, however, any inequality between [Cl$^-$]$_i$ and [Cl$^-$]$_o$ appears as a voltage across the plasmalemma. The magnitude of this membrane voltage or potential (E) may be calculated by considering the electrical work (Q$_e$) in coulomb volts or joules that would be obtainable by allowing one mole of Cl$^-$ to move down the voltage gradient, from C$_i$ to C$_o$. This is equivalent to treating the membrane as a leaky capacitance:

$$Q_e = zFE \tag{1.8}$$

where z is the valency of chloride (-1) and F is **Faraday's constant**, with a value of approximately 96 500 coulombs mole^{-1}.

Equation 1.8 dealt only with the voltage gradient and equation 1.7 dealt only with the solute concentration gradient. For a reversible system at equilibrium, however, the mechanical work (Q$_c$) carried out by the piston must be equal to the electrical work (Q$_e$) obtainable from the system. Hence:

$$Q_e = Q_c$$

It follows from equations 1.7 and 1.8 that:

$$zFE = RT ln \frac{[Cl^-]_o}{[Cl^-]_i}$$

$$\text{or} \quad E = \frac{RT}{zF} ln \frac{[Cl^-]_o}{[Cl^-]_i} \tag{1.9}$$

This is known as **Nernst's equation**, after its originator, the German physical chemist W.H.Nernst (1864–1941). One may refer to E under these circumstances either as the **membrane potential** or as the **electrochemical diffusion potential** for Cl^-. Equation 1.9 is simultaneously applicable to all concentration cells operating across the membrane, provided that all the ions involved are permeant. This is approximately true in most cells for both Cl^- and K^+. Hence:

$$E = \frac{RT}{z_{Cl}F} ln \frac{[Cl^-]_o}{[Cl^-]_i} = \frac{RT}{z_K F} ln \frac{[K^+]_o}{[K^+]_i} \simeq -90\,mV$$

$$\text{where} \quad z_{Cl} = -1 \quad \text{and} \quad z_K = +1$$

Hence: $-ln \dfrac{[Cl^-]_o}{[Cl^-]_i} = ln \dfrac{[K^+]_o}{[K^+]_i}$

Thus: $\dfrac{[Cl^-]_o}{[Cl^-]_i} = \dfrac{[K^+]_i}{[K^+]_o}$

$$\text{or} \quad [K^+]_o[Cl^-]_o = [K^+]_i[Cl^-]_i \tag{1.10}$$

where $[Cl^-]_o > [Cl^-]_i$, as shown in equation 1.3.

This is known as a **Donnan equilibrium**, after the British physical chemist who first drew attention to its wide application in biological cells (Donnan, 1924). The thermodynamic basis of this type of equilibrium, however, was studied many years earlier by the American mathematical physicist J.W.Gibbs (1839–1903). The situation is sometimes referred to as a **Gibbs–Donnan equilibrium** to denote this fact. One may restate equation 1.10 in more general terms as:

$$[K^+]_o[Cl^-]_o \ldots \ldots [\text{permeant ion}]_o$$
$$= [K^+]_i[Cl^-]_i \ldots \ldots [\text{permeant ion}]_i \tag{1.11}$$

Any participating multivalent ion will appear in equation 1.11 as an appropriate root of the concentration value. Thus, any divalent calcium ions (Ca^{2+}) would appear as:

$$[K^+]_o[Cl^-]_o\sqrt[2]{[Ca^{2+}]_o} = [K^+]_i[Cl^-]_i\sqrt[2]{[Ca^{2+}]_i}$$

– 2

Osmotic considerations

Equation 1.2 shows that:

$$[K^+]_i = [Cl^-]_i + [Pr^-]_i$$

By adding $[K^+]_i$ to both sides of this equation we obtain:

$$2[K^+]_i = [K^+]_i + [Cl^-]_i + [Pr^-]_i \tag{2.1}$$

Likewise, equation 1.2 shows that:

$$[K^+]_o = [Cl^-]_o$$

By adding $[K^+]_o$ to both sides of the latter equation we obtain:

$$2[K^+]_o = [K^+]_o + [Cl^-]_o \tag{2.2}$$

The right-hand sides of equations 2.1 and 2.2 represent the total osmolarity in C_i and C_o respectively. Since equation 1.10 shows that $[K^+]_i > [K^+]_o$, it follows that:

$$[K^+]_i + [Cl^-]_i + [Pr^-]_i > [K^+]_o + [Cl^-]_o$$

Hence:

$$\text{osmolarity in } C_i > \text{osmolarity in } C_o \tag{2.3}$$

The plasmalemma being water permeable, it follows that the cell would swell, or even burst, as water enters. The osmotic problem is overcome in many cells by an unequal distribution of sodium ions (Na^+) across the plasmalemma. Extracellular fluids contain abundant Na^+, whereas intracellular fluids have a much lower $[Na^+]$, thereby rebalancing the osmolarities in equation 2.3. A cation pump in the plasmalemma, which is fuelled by adenosine triphosphate, extrudes Na^+ and keeps $[Na^+]_i$ lower than $[Na^+]_o$. The normally low permeability of the plasmalemma to Na^+ aids in the preservation of this asymmetry. Failure of the Na pump, or any substantial entry of Na^+, however, will lead to cellular swelling. This is a common sign of cellular injury. The same ion pump that extrudes Na^+ also moves K^+ in the inward direction, and is spoken of as the

Na/K pump for this reason. Opposing ion fluxes are obligatorily linked, with approximately three Na^+ ions moved outwards for each pair of K^+ ions moved inwards, although the ratio differs somewhat in different cells. The inequality ensures that the pump preserves osmotic balance across the plasmalemma, which it would not do if the fluxes of Na^+ and K^+ were equal. The inequality also creates a current (e.g. $3Na^+ - 2K^+$) and the pump is said to be electrogenic or rheogenic for this reason. The immediate contribution which the pump makes to membrane potential, however, is usually only between 1 and 10 mV, which is to say between 1 and 10% of the total. This is demonstrated by the initially quite small reduction in membrane potential which usually results from drug-induced inhibition of the pump. The reader should note that unequal $[Na^+]$ on the two sides of the plasmalemma itself creates a Gibbs–Donnan type of redistribution among more permeant ions, similar to that created by unequal $[protein^-]$ distribution (Chapter 1). Moreover, the asymmetry of distribution among more permeant ions which results from the unequal $[Na^+]$ is very similar to that due to the unequal $[protein^-]$. In both cases, at equilibrium, there will be $[Cl^-]_i < [Cl^-]_o$ and $[K^+]_i > [K^+]_o$. For this reason, therefore, living tissues are said to display a double Gibbs–Donnan type of equilibrium.

The asymmetric distribution of Na^+ across the plasmalemma has electrical consequences, as well as the osmotic consequences mentioned above. These electrical consequences stem from the fact that the electrochemical diffusion potential for Na^+ will have an opposite polarity to that for K^+ or Cl^- (Chapter 3), with the inside positive with respect to the outside.

- 3

Ions to which the plasmalemma has limited permeability

Equations 1.9 and 1.10 relate to a situation in which the plasmalemma is permeable to all specified ions and the electrochemical diffusion potentials are identical in both magnitude and polarity. If the diffusion potentials are of opposing polarity, however, a different set of equations applies. These may be derived from the equivalent electrical circuit shown in Figure 3.1. Here, two compartments (inside and outside) are separated by a membrane generating two electrochemical diffusion potentials (E_{Na} and E_K), represented as conventional voltage sources of opposing polarity. This opposition arises because $[Na^+]_o > [Na^+]_i$ and $[K^+]_o < [K^+]_i$ in most mammalian cells, for the reasons discussed in Chapters 1 and 2 (equations 1.9, 1.10 and 2.3). Both cations are assumed to permeate the plasmalemma exclusively via their respective ion-selective channels, the electrical resistances of which are designated R_{Na} and R_K. Membrane capacitance (f_m) and transmembrane potential (E) complete the circuit, with the latter conventionally expressed as inside with respect to outside. Consequently, E has a negative value in most mammalian cells because the value and sign of E are largely determined by E_K and not by E_{Na}. Current (I) flows around the circuit in the direction shown by the arrows. It is carried by K^+ via R_K and by Na^+ via R_{Na}.

Ohm's law, which states that E = IR (where R is resistance in ohms if E is in volts and I is in amperes), may be applied in Figure 3.1 to that part of the circuit (R_{Na} and E_{Na}) known as the Na^+ channel to give:

$$E_{Na} - E = IR_{Na}$$

Similarly, if Ohm's law is applied to the K^+ channel part of the circuit it gives:

$$E - E_K = IR_K$$

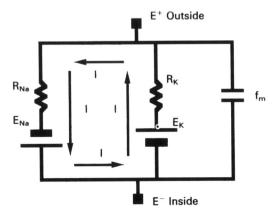

Figure 3.1 The equivalent electrical circuit for a membrane separating two compartments (inside and outside). Transmembrane potential (E) is a resultant of two ion channel generators of opposing polarity. The Na$^+$ channel generator consists of an electrochemical diffusion potential (E$_{Na}$) operating via a resistance (R$_{Na}$). The K$^+$ channel generator consists of a different electrochemical diffusion potential (E$_K$), operating via a different resistance (R$_K$). A current (I) flows around the circuit in the direction of the arrows. Membrane capacitance is f$_m$, defined as amount of ionic charge stored per unit of transmembrane voltage.

Since I must have the same value at all points in the circuit, it follows that:

$$\frac{E_{Na} - E}{E - E_K} = \frac{R_{Na}}{R_K} = \frac{g_K}{g_{Na}}$$

where g is the conductance in siemens, or units of reciprocal ohmic resistance. (Note that conductance will increase as I increases, and is, therefore, more covenient to use than resistance, which will decrease as I increases.)

If R$_K$ is small with respect to R$_{Na}$, as it is in most mammalian cells, the value of E − E$_K$ also will be small and E $\simeq$ E$_K$. In consequence, the value of I will be small, and equations 1.9 and 1.10 will still apply as a first approximation. If both R$_{Na}$ and R$_K$ are small then I will be substantial, with E adopting a value intermediate between E$_K$ and E$_{Na}$. Where R$_{Na}$ is small with respect to R$_K$, a situation which occurs transiently in some mammalian cells (Chapter 4), the value of I becomes small and E $\simeq$ E$_{Na}$. This means that equations 1.9 and 1.10 again apply, but with E$_{Na}$ now substituted for E$_K$ (or E$_{Cl}$). Note that a large value of I represents a large entry of Na$^+$, and a consequential large loss of K$^+$ or entry of Cl$^-$ so that the ion gradients will

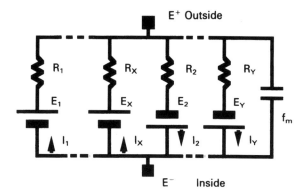

Figure 3.2 The equivalent electrical circuit for a membrane separating two compartments (inside and outside). Transmembrane potential (E) is a resultant of two groups of ion channel generators of opposing polarity. Electrochemical diffusion potentials E_1 to E_x drive the outside of the membrane to more positive potential, whereas E_2 to E_y drive it to less positive, or more negative, potential. Thus, E_2 to E_y tend to diminish membrane polarity (which is outside positive), whereas E_1 to E_x tend to enhance it. The former are said to depolarize, and the latter to hyperpolarize the cell membrane. Resistances of ion channels are labelled R. Current (I) flows through each ion channel in the direction of the arrow.

gradually dissipate. It is the function of the Na/K pump to preserve these gradients over the long term. The short-term dissipation of the gradients will create the transmembrane currents upon which electrical activity of mammalian cells depends. The dissipation arises because $[Na^+]_o > [Na^+]_i$ and because R_{Na} is less than infinite. Under physiological conditions it is the variation in R_{Na} which is largely responsible for the variation in I.

Figure 3.1 contains two ion channels, but may be generalized in the way shown in Figure 3.2. In the latter circuit, the sum of all inward currents must equal the sum of all outward currents, hence:

$$I_1 + \ldots\ldots I_x = I_2 + \ldots\ldots\ldots\ldots I_y \tag{3.1}$$

If E_c represents the electrochemical diffusion potential for a particular ion species in Figure 3.2, the concentration of which differs between inside and outside, it follows from Ohm's law that:

$$E - E_c = I_c R_c = \frac{I_c}{g_c}$$

where R_c and g_c are the resistance and conductance respectively of the ion channel through which the current (I_c) is flowing.

This may be rewritten as:

$$I_c = g_c(E - E_c)$$

This permits a substitution for I in equation 3.1, giving:

$$g_1(E - E_1) + \ldots \ldots g_x(E - E_x)$$
$$= g_2(E_2 - E) + \ldots \ldots g_y(E_y - E)$$

or $g_1E - g_1E_1 + \ldots \ldots g_xE - g_xE_x$
$$= g_2E_2 - g_2E + \ldots \ldots g_yE_y - g_yE$$

or $g_1E + g_xE + \ldots \ldots g_2E + g_yE$
$$= g_2E_2 + g_yE_y + \ldots \ldots g_1E_1 + g_xE_x$$

or $E(g_1 + g_x + \ldots \ldots g_2 + g_y)$
$$= g_2E_2 + g_yE_y + \ldots \ldots g_1E_1 + g_xE_x$$

where $(g_1 + g_x + \ldots \ldots g_2 + g_y)$ represents the total ionic conductance (g_T) of the membrane. Thus:

$$E = \frac{g_2}{g_T}(E_2) + \frac{g_y}{g_T}(E_y) + \ldots \frac{g_1}{g_T}(E_1) + \frac{g_x}{g_T}(E_x) \tag{3.2}$$

This is often termed the **chord conductance equation**. Equation 3.2 shows that the membrane potential (E) is equal to the algebraic sum of a series of terms, one for each of the contributing ion channels. Each term is the product of an electrochemical diffusion potential (E_c) and a ratio of two conductance values, the numerator being the conductance of the specific ion channel (g_c) and the denominator the total ionic conductance of the membrane (g_T). The contribution that each ion channel makes to the value of E, therefore, depends upon two things. Firstly, upon the magnitude and sign of E_c and, secondly, upon its conductance (g_c). The greater the value of the latter relative to g_T, the more closely will E resemble the relevant E_c value. If only one ion channel has substantial conductance, that one will approximate to g_T and for that ion channel g_c/g_T becomes approximately unity. Only one term of significant magnitude remains on the right-hand side of equation 3.2, which simplifies to a version of equation 1.9 in which E approximates to the E_c value of the permeant ion channel ($E_c = E_{Cl} = E$ in equation 1.9).

Equations 3.1 and 3.2 show that the amount of current flowing through a particular ion channel in Figure 3.2 will be governed by the difference between the membrane potential (E) and the electrochemical diffusion potential (E_c) for that channel. In those ion channels (E_1 to E_x) which have the same polarity as E, that is to say, inside negative, current flows outwards. The reverse is true for channels E_2 to E_y. Equation 3.2 shows that the overall value of E is the resultant of, and thus intermediate between, a series of negative E_c values (E_1 to E_x) and a series of opposing positive E_c values (E_2 to E_y). The contribution of the latter group is smaller than that of the former group, primarily because members of the latter group display lower g_c values than members of the former group. Any increase in

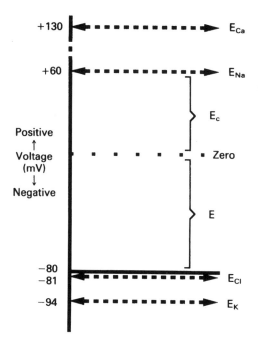

Figure 3.3 Electrochemical diffusion potentials (horizontal dashed lines) for Ca^{2+}, Na^+, Cl^- and K^+ in a typical mammalian cardiac muscle cell, expressed as voltage inside the cell with respect to outside. For comparison, a typical value of resting membrane potential (E) is indicated by the continuous horizontal line at -80 mV. Note how closely E conforms to E_{Cl}, although it is somewhat positive with respect to E_K. Note also that both E_{Ca} and E_{Na} are substantially positive with respect to E. Net transmembrane force on an ion is the algebraic difference between the value of E and the electrochemical diffusion potential (E_c) for that ion, and has the value E minus E_c (which is -140 mV for Na^+, as illustrated). Negative values of net transmembrane force signify that force operates inwards, as in the case of Ca^{2+} and Na^+. Positive values of net transmembrane force indicate that force operates outwards, as in the case of K^+. This means that for a cell at seeming equilibrium there is a continuous outward K^+ current (I_K) and continuous inward Na^+ and Ca^{2+} currents (I_{Na} and I_{Ca}). These ion fluxes have to be exactly counterbalanced by energy-consuming ion pumps if the ion concentration gradients are to be preserved. In contrast, there is no net flux of Cl^- at equilibrium because of the absence of a net transmembrane force on Cl^-. The distribution of Cl^- is said to be passive for this reason. This means that the distribution of $[Cl^-]$ is determined by the concentration gradients of the other ions, and thus the membrane voltage, and there is no chloride pump to disturb the equilibrium.

conductance of ion channels E_2 to E_y will diminish the transmembrane voltage E. This is spoken of as **depolarization**. In contrast, **hyperpolarization** occurs as a result of any increase in g_1 to g_x. Increased total outward current through E_1 to E_x produced in this latter way will be associated with an exactly equal increase in total inward current through E_2 to E_y. This would occur because the hyperpolarized membrane is now at a voltage which is even further removed from the E_c values of channels 2 to y than previously.

Consider the situation where only two ion channels contribute to equation 3.2, one of each polarity, say, E_K and E_{Na}. Equation 3.2 may then be written as:

$$E = \frac{g_K}{g_K + g_{Na}} E_K + \frac{g_{Na}}{g_K + g_{Na}} E_{Na}$$

where E and E_K have negative values, but E_{Na} is positive.

Since usually $g_K \gg g_{Na}$ the $g_K/g_K + g_{Na}$ term approximates to unity and the $g_{Na}/g_K + g_{Na}$ term is small, so E approximates to E_K, but remains a little more positive than E_K. The depolarizing influence of a given increase in g_{Na} will depend upon the prevailing value of g_K. This is because the decrease in the $\dfrac{g_K}{g_K + g_{Na}}$ term for a given increase in g_{Na} will be greater the smaller the value of g_K, just as the increase in the $\dfrac{g_{Na}}{g_K + g_{Na}}$ term will be greater the smaller the value of g_K.

Figure 3.3 shows the E_c values for several naturally occurring ions, together with the membrane potential for a typical resting cardiac muscle cell. Such cells have low g_{Na} and g_{Ca} values. When electrically stimulated during systole, however, they demonstrate large but temporary increases in both g_{Na} and g_{Ca} values. Such cells are said to possess electrical excitability and they generate so-called action potentials. When g_{Na} becomes the major ionic conductance of the membrane, the value of E in equation 1.9 approximates to E_{Na}; that is to say, to an inside positive voltage of $+60\,\text{mV}$. This reversal of membrane polarity is dealt with in Chapter 4.

Readers who desire to have a fully referenced historical account of the development of ideas about the material that is outlined in Chapters 1, 2 and 3 should consult the masterly review prepared by Sperelakis (1979).

– 4

Electrically excitable cells

Electrical stimuli may be conveniently delivered to biological cells by means of metallic electrodes through which current is passed. Alternatively, electrolyte-filled hollow glass tubes may be used. The latter are particularly useful because they may be made very narrow and then inserted without injury into the interior of cells via the plasmalemma. The punctured plasmalemma reseals around the tube. If a negative electrode is placed on the outside of the plasmalemma, and an anode is connected to the cell interior, the passage of current between the electrodes causes depolarization or a loss of transmembrane potential. The magnitude of the depolarization is dependent upon current strength (I). A small local depolarization, known as an electrotonic potential, is produced by a suitably small current, as shown by line **a** in Figure 4.1. The depolarization produced by the current is not instantaneous, however, due to the time taken for electrical capacitance in the membrane to charge or discharge (Figures 3.1 and 3.2). Larger currents induce larger and more rapidly developing electrotonic potentials (Figure 4.1, lines **a** and **b**). In certain cell types, notably those in nerve and muscle, still larger currents cause the initial electrotonic potential to transform into an action potential (Figure 4.1, line c). The transition occurs after a delay called the **electrotonic latent period** and at a membrane potential known as the **threshold voltage**. If stimulant current ceases before threshold voltage is reached, the electrotonic potential already attained will immediately start to decay, as shown by line **d** in Figure 4.1.

An action potential differs from an electrotonic potential in several respects. For one thing, the magnitude of the latter declines with increasing distance from the stimulating electrodes (Figure 4.2), but it does not decline spontaneously with the passage of time, provided that stimulating current continues to flow. The results shown in Figure 4.2 are from an experiment in which subthreshold stimulating current had continued for sufficient time for an electrotonic equilibrium or a plateau to be reached, as shown at the right-hand end of lines a or b in Figure 4.1.

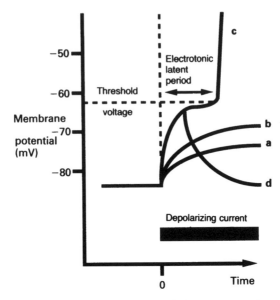

Figure 4.1 Time course of membrane potential changes in an electrically excitable cell through which depolarizing current (I) was passed, producing an electrotonic depolarization, starting at time zero. The value of I was greater in line b than in line a, but subthreshold in both cases. In line c, however, threshold voltage was reached after a delay known as the electrotonic latent period. If I ceased before threshold was reached the electrotonic depolarization dissipates (line d).

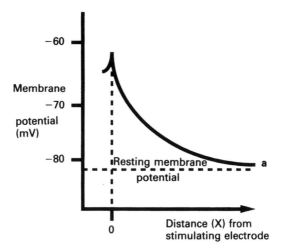

Figure 4.2 Spatial changes in membrane potential during an electrotonic depolarization recorded at varying distances (X) from the point of application (zero) of a small constant depolarizing current. Sufficient time was allowed for maximal depolarization to be reached, as at the right-hand end of line **a** in Figure 4.1.

Action potentials, unlike electrotonic potentials, involve transient depolarizations of the membrane. Their duration is independent of the time for which stimulant current continues to flow after threshold voltage has been reached. Moreover, once an action potential is initiated, it spreads out automatically until it reaches the tissue margins. It cannot remain stationary because it brings to each site that it invades an intrinsic, time-dependent inactivating mechanism.

Figure 4.3a represents an equivalent electrical circuit for electrotonic depolarization in a long fibre of uniform circular cross-section. Two chains of resistors are shown, representing longitudinal resistance of extra- and intracellular compartments respectively. These two chains interconnect across the membrane at regular intervals via ion-conducting channels with a resistance of R_m. When the switch is closed the electrical stimulator (on the left) permits the extracellular electrode (S^-) to become the cathode and an intracellular electrode (S^+) to become the anode. At A the membrane depolarizes by adopting a voltage somewhere between that of the stimulating electrode (outside negative) and that of the unstimulated membrane (outside positive). In Figure 4.3b, on the other hand, both electrodes are extracellular. Electrotonic depolarization at B in Figure 4.3b would be similar to that at B in Figure 4.3a, whereas at X and Y in Figure 4.3b there would be electrotonic hyperpolarization. If the polarity of the electrodes in Figure 4.3a is reversed, with the anode outside and the cathode inside, hyperpolarization will occur at points A to C.

Whenever the membrane potential gradient of a cell is more than fully opposed by the diffusional force created by an existing ion concentration gradient, the relevant ion diffuses through the membrane in the direction of the concentration gradient, and against the voltage gradient. This occurs because the values of Q_c and Q_e in equations 1.7 and 1.8 are no longer equal. The flow of ions constitutes an electric current. The larger the discrepancy between Q_c and Q_e the larger will be the current. If the membrane potential gradient of a cell is greater than the diffusional force created by an existing concentration gradient, on the other hand, the relevant ion migrates in the direction of the voltage gradient, and towards the region of already higher concentration, thereby adding to it. In Figure 4.3a, therefore, outward current (I_A) will occur at A. The current will be carried by K^+ and/or Cl^- in the membranous parts of the circuit. It should be noted that although Cl^- moves inwards, this constitutes an outward current, since electrical current is treated as a flow of positive charge. This is similar to the convention of treating

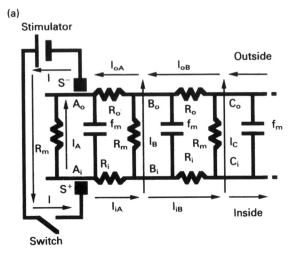

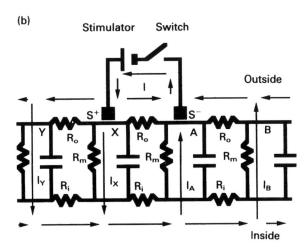

Figure 4.3a–b Equivalent electrical circuits for electrotonic changes in plasmalemmal voltage of a uniform muscle fibre, the internal longitudinal resistance of which is R_i and the transmembrane resistance and capacitance of which are R_m and f_m respectively. Resistance of the external bathing fluid is R_o. Current (I) is delivered to the fibre from a stimulator via a pair of electrodes (S). In 4.3a, one electrode is placed on either side of the plasmalemma and an electrotonic depolarization occurs when the switch is closed. In 4.3b, both electrodes are placed on the outside of the plasmalemma, and a mixed electrotonic response occurs when the switch is closed, with depolarization spreading out from S^- and hyperpolarization from S^+.

the current in a metallic wire as flowing from the anodic to the cathodic end. In reality, of course, it is a flow of negatively charged electrons through the wire in the reverse direction.

4.1 CABLE PROPERTIES

The electrode S^- at A_o may be thought of as drawing current from the membrane, and the electrode S^+ at A_i as delivering current to the membrane. Some of this current passes directly through the membrane between A_i and A_o, but some of it passes axially right-wards via the intracellular resistance chain (R_i). The magnitude of this rightward axial current (I_i) diminishes progressively as it is lost through the membrane to the extracellular fluid. The leftward movement of I_o signifies that B_o must be at a somewhat higher potential than A_o, just as the rightward movement of I_i shows that A_i is at a higher potential than B_i. If B_i is at a more negative (or less positive) potential than A_i, and B_o is at a more positive potential than A_o, the transmembrane voltage gradient at B (i.e. B_i to B_o) must be more polarized (i.e. less depolarized) than the transmembrane voltage gradient at A (i.e. A_i to A_o). In other words, electrotonic depolarization becomes less marked the further rightwards one goes from the stimulating electrodes.

The transmembrane voltage (E), conventionally expressed as inside with respect to outside, may now be calculated at various places along the fibre depicted in Figure 4.3a. Extracellular fluid usually has a very low electrical resistance. As a first approximation, therefore, one may treat all points on the external surface of the plasmalemma as being at a uniform potential. Transmembrane potential thus approximates to the measured internal potential. In most experiments the extracellular fluid is earthed. The transmembrane current (I_m) that flows at a point X located at a distance x from the stimulating electrodes is defined as:

$$I_m = -\frac{dI_i}{dx} \tag{4.1}$$

since the decline in I_i with distance constitutes I_m.

Axial current at point X may be defined from Ohm's law as:

$$I_i = -\frac{dE}{dx} \cdot \frac{1}{R_i} \tag{4.2}$$

where E is the measured membrane potential at point X.

Combining equations 4.1 and 4.2 gives:

$$I_m = \frac{d^2E}{dx^2} \cdot \frac{1}{R_i} \tag{4.3}$$

This is the most general form of the **electrical cable equation**.

When the switch in Figure 4.3a initially closes, some of the transmembrane current (I_m) is used to alter the charge stored in the membrane capacitors (F_m). The remainder (I_R) flows resistively to the exterior of the cell via R_m. Capacitative current (I_F) is related to resistive current (I_R), and to I_m, as follows:

$$I_m = I_R + I_F \tag{4.4}$$

The capacitative current I_F (in amperes) is the charge (in coulombs) moving per second and can be defined as:

$$I_F = F_m \cdot \frac{dE}{dt} \tag{4.5}$$

where t is the time in seconds that has elapsed since closing the switch, and F_m is the membrane **electrical capacity** (in farads) or charge (in coulombs) stored per unit of transmembrane voltage (E).

Furthermore, the resistive current (I_R) is defined by Ohm's law as:

$$I_R = \frac{E}{R_m} \tag{4.6}$$

Combining equations 4.3–4.6 gives a useful form of the general electrical cable equation:

$$\frac{d^2E}{dx^2} \cdot \frac{1}{R_i} = \frac{E}{R_m} + F_m \cdot \frac{dE}{dt} \tag{4.7}$$

The left-hand term relates to changes of E at constant t values and the term at the right-hand end relates to changes of E at constant x values.

Sometimes this is more conveniently arranged to give:

$$E = R_m \cdot F_m \frac{dE}{dt} - \frac{R_m}{R_i} \cdot \frac{d^2E}{dx^2}$$

where E is the transmembrane voltage at time t and distance x.

When an electrotonic potential has reached its plateau value (e.g. at the right-hand end of line b in Figure 4.1), all capacitative current

(I_F) will have ceased because the capacitors are fully charged. So equation 4.7 simplifies to:

$$\frac{d^2E}{dx^2} \cdot \frac{1}{R_i} = \frac{E}{R_m} \quad \text{or} \quad E = \frac{R_m}{R_i} \cdot \frac{d^2E}{dx^2} \tag{4.8}$$

Resistances R_m and R_i in this and the preceding equations relate to a unit length of fibre. For a long, uniform cylinder of circular cross-section and radius a it is more convenient to use resistivities (r), which relate to a unit area of the plasmalemma (r_m) or to a unit cross-sectional area of the fibre (r_i), where:

$$R_m = \frac{r_m}{2\pi a} \quad \text{and} \quad R_i = \frac{r_i}{\pi a^2}$$

This permits equation 4.8 to be rewritten with resistivities in place of resistances:

$$E = \frac{ar_m}{2r_i} \cdot \frac{d^2E}{dx^2} \tag{4.9}$$

Equation 4.8 requires a mathematical transform in order to be solved for E. (For suitable Laplace or Fourier transforms see Hodgkin and Rushton (1946)). The solution is:

$$\frac{E}{E_o} = e^{\frac{-x}{\sqrt{(R_m/R_i)}}} \tag{4.10}$$

where E_o is the value of E at point A in Figure 4.3a, so x = 0, and e is the exponent, or the **base of natural logarithms**, with a value of 2.72 approximately.

Equation 4.10 shows that E declines exponentially with increasing values of x. The value of $\sqrt{(R_m/R_i)}$ is called the **space constant**. If resistivities are used instead of resistances, however, the space constant becomes $\sqrt{(ar_m/2r_i)}$. This indicates that the space constant is proportional to $\sqrt{a}$. Other things being equal, therefore, electrotonic potentials will extend further from a stimulating electrode in a larger diameter fibre than in a smaller diameter fibre. Electrotonic potentials also extend further when R_m is high or when R_i is low. This has important functional implications which are explored in later chapters of the book. The space constant may be thought of as the value of x at which

$$E = \frac{E_o}{e} \tag{4.11}$$

Suppose now that the fibre shown in Figure 4.3a is depolarized uniformly along its length instead of in a graded manner. (This is only possible in most mammalian cells over short distances, and may be quite difficult to arrange experimentally over distances in excess of 1–2 mm). E will not change with x, and the left-hand side of equation 4.7 will be zero. Hence:

$$0 = \frac{E}{R_m} + F_m \cdot \frac{dE}{dt}$$

This rearranges to give:

$$E = -R_m \cdot F_m \cdot \frac{dE}{dt}$$

and may be solved by integration to give:

$$\frac{E}{E_\infty} = 1 - e^{\frac{-t}{R_m \cdot F_m}} \tag{4.12}$$

where E_∞ is the plateau value of E, and corresponds to the value at the right-hand end of line b in Figure 4.1 ($t \simeq \infty$).

This value of $R_m \cdot F_m$ is known as the **time constant** (τ) and represents the value of t at which:

$$\frac{E}{E_\infty} = 1 - \frac{1}{e} \tag{4.13}$$

So far, membrane capacity (F_m) has been expressed per unit length of the fibre. Capacitance (f_m), (expressed per unit area of plasmalemma) may be used instead. The relationship between them is:

$$f_m = \frac{F_m}{2\pi a}$$

This permits the time constant ($R_m \cdot F_m$) to be rewritten using resistivity (r_m) and capacitance (f_m) as follows:

$$F_m \cdot R_m = \frac{2\pi a \cdot f_m}{1} \cdot \frac{r_m}{2\pi a} = f_m \cdot r_m$$

where a is the fibre radius.

This means that the time constant, unlike the space constant, is independent of fibre radius.

Equation 4.3 expressed total transmembrane current (I_m) in a unit length of the fibre. Current density (i_m), passing through a unit area of plasmalemma, is related to I_m as follows:

$$i_m = \frac{I_m}{2\pi a}$$

If membrane current density (i_m) is substituted for I_m in equation 4.3, we obtain:

$$i_m = \frac{d^2E}{dx^2} \cdot \frac{1}{2\pi a \cdot R_i}$$

If resistivity (r_i) is substituted for R_i in the above equation we get:

$$i_m = \frac{d^2E}{dx^2} \cdot \frac{\pi a^2}{2\pi a \cdot r_i} = \frac{d^2E}{dx^2} \cdot \frac{a}{2r_i} \tag{4.14}$$

This is the most generally useful form of the electrical cable equation.

4.2 CONDUCTED ACTION POTENTIALS

In most cells, during an action potential, there is a sudden increase in g_{Na}, with a consequent reversal of the polarity of membrane voltage, as mentioned in connection with Figure 3.3. Thus, the inside becomes positive with respect to the outside. The equivalent electrical circuit is shown in Figure 4.4. This is a modified version of Figure 4.3a, with the 'stimulator' (at S) incorporated into the membrane. The stimulator is still represented by a conventional symbol for a voltage source, albeit a biological one now. This induces a current (I) to flow inwards across the membrane. The current is carried by Na^+, as discussed in Chapter 3. In Figure 4.4, however, an action potential, and not an externally supplied current, causes the electrotonic depolarization at distant membrane sites. Nevertheless, in both Figures 4.3a and 4.4, $I_A > I_B$. Just as the membrane is depolarized electrotonically more at A than at B in Figure 4.3a, so also in Figure 4.4. In so far as an electrotonic potential at A or B in Figure 4.4 can take the membrane there to threshold voltage, the action potential already at S will be extended rightwards. That is to say, it conducts or begins again at A or B. It takes a finite time, however, for I_A or I_B to discharge the intervening capacitors, so threshold will not be reached simultaneously at A and B. The larger the space constant, the further to the right of S will points on the membrane reach threshold as a result of the action potential already at S.

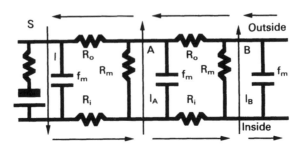

Figure 4.4 Conducted action potential. An equivalent electrical circuit for electrotonic depolarization of the plasmalemma of a uniform muscle fibre, the internal longitudinal resistance of which is R_i and the transmembrane resistance and capacitance of which are R_m and f_m respectively. Resistance of the external bathing fluid is R_o. Inward current (I) occurs at S due to the presence of the early phase of an action potential, represented for convenience by a conventional symbol for a voltage source. The circuit is analogous to Figure 4.3a, except that here the 'stimulator' is an internally generated action potential at S.

Conduction velocity (θ) of an action potential in a uniform fibre is defined as:

$$\theta = \frac{x}{t} \quad \text{or} \quad x = \theta t$$

where x is the distance that an action potential travels in time t.

Hodgkin (1954) pointed out that θ may also be expressed as a differential equation involving E that takes the form:

$$\frac{dE}{dx} = \frac{1}{\theta} \cdot \frac{dE}{dt} \quad \text{or} \quad \frac{d^2E}{dx^2} = \frac{1}{\theta^2} \cdot \frac{d^2E}{dt^2} \tag{4.15}$$

For the particular value of i_m that takes the membrane potential to threshold, so that a conducted action potential occurs, the value of the d^2E/dx^2 term in equation 4.15 may be substituted in equation 4.14 to give:

$$i_m = \frac{a}{2\theta^2 \cdot r_i} \cdot \frac{d^2E}{dt^2}$$

This may be rearranged to give:

$$\theta^2 = \frac{a}{2i_m \cdot r_i} \cdot \frac{d^2E}{dt^2} \quad \text{or} \quad \theta = \sqrt{\left(\frac{a}{2i_m \cdot r_i} \cdot \frac{d^2E}{dt^2}\right)} \tag{4.16}$$

Equation 4.16 shows that the conduction velocity of an action potential (θ) is directly proportional to $\sqrt{a}$, and inversely proportional to $\sqrt{r_i}$, as was the case for the space constant in equation 4.10. Thus, action potentials conduct fastest in wide fibres having a low longitudinal resistance. Equation 4.16 also shows that θ is dependent upon how membrane potential (E) changes with time (t) during the action potential (d^2E/dt^2). The membrane potential during an action potential changes most abruptly at the start. It is this early phase of the action potential, therefore, which determines its rate of conduction. Most injured cells are partially depolarized and generate slowly rising action potentials, for reasons that will be dealt with in Chapter 5 (Figure 5.4). Slowly rising action potentials conduct slowly. This is of significance in certain disturbances of rhythm in injured hearts (Chapter 9).

4.3 CELL-TO-CELL CONDUCTION

Action potentials not only spread along the surface of an excitable cell, but may also invade closely adjacent cells. This is particularly important in the heart, where specialized aqueous, cell-to-cell connections are present in the regions of closest intercellular contact. These are called **intercalated discs** (Figure 4.5b). These discs are analogous to the gap junctions or nexuses that are to be found in many mammalian tissues. Each gap junction contains a collection of hollow, protein-lined cylinders called **connexons**, seen best in freeze-fracture electron micrographs. The aqueous lumen of each connexon provides a path of low electrical resistance between the cytoplasm of cells that it links together. The frequency with which connexons are found in different parts of the cardiac plasmalemma determines electrical resistance (R_i) and resistivity (r_i) in equations 4.2–4.14, and hence the tissue electrotonic space constant $[\sqrt{(R_m/R_i)}]$. Intercalated discs are more abundant in end-to-end connections between cardiac myocytes than in side-to-side connections, so R_i is minimized in the longitudinal direction, and the space constant consequently maximized. Action potential conduction velocity through the myocardium is greater, therefore, in the direction of the main fibre bundles. The older and more fibrotic (and thus insulated) myocardial fibres become the more anisotropic becomes their con-duction velocity. This has considerable significance for disturbances in cardiac rhythm (Chapter 9).

The ionic composition of the cytoplasm has a considerable influence upon the permeability of connexons to cytoplasmic solutes of all types. Any persistent rise in $[Ca^{2+}]_i$ or $[H^+]_i$, such as occurs in most types of cellular injury, for example, closes the connexons and

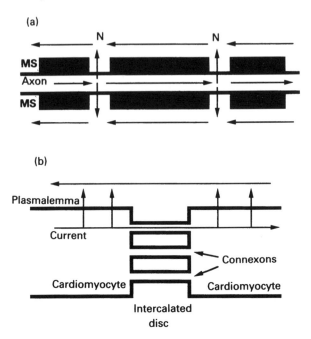

Figure 4.5a–b Electrotonic flow of current in two non-uniform cable structures. In 4.5a, an interrupted myelin sheath (**MS**) surrounds and electrically insulates a nerve axon. Interruptions in the MS, called **nodes of Ranvier**, allow longitudinal cytoplasmic current (analogous to I_i in Figure 4.3a) to escape at intervals to the extracellular fluid. In 4.5b, two cardiomyocytes are linked via an intercalated disc containing connexons. Intracellular longitudinal current (analogous to I_i in Figure 4.3a) flows from one myocyte to the other via the connexons, but to extracellular fluid only via the plasmalemma.

raises their electrical resistance. It is possible that this protects cells adjacent to an injured one, by limiting what would otherwise be a far-reaching leak of valuable cytoplasmic constituents, such as enzymes, to the extracellular space via the injured cell. This protection is obtained, however, at the price of a reduction in action potential conduction velocity in the area, due to the rise in r_i. Slowing of conduction of action potentials in injured areas of myocardium is associated with important disturbances of rhythm (Chapter 9). The speed of action potential conduction is increased by sympathetic nervous stimulation of the myocardium, which raises the level of cyclic adenosine monophosphate in the cytoplasm, causing phosphorylation of the connexons, thereby raising their ionic conductance. This is particularly important in the atrioventricular node (Chapter 9).

Wherever some of the current flowing rightwards inside the tissue in Figure 4.3a escapes across the plasmalemma to the extracellular compartment, the remaining intracellular axial current will decrease. The more widely spaced are the points where transmembrane escape of current occurs, the further will the current and electrotonic potentials spread, and the greater will be the calculated space constant. This is particularly marked in myelinated nerve fibre axons, as shown in Figure 4.5a, where the only regions available for transmembrane escape of resistive current are at the nodes (N), the intervening axon segments being electrically insulated from the extracellular fluid by a thick myelin sheath (MS). The myelin-sheathed region still permits the axon to display a capacitance, and this can absorb some charge from the axon current (equations 4.4–4.7). Other things being equal, however, electrotonic depolarization in a myelinated axon extends further from the point of application of a given amount of stimulant current than in an unmyelinated axon. The lines of arrows in Figure 4.5a represent the flow of such current. Stimulation is from the left-hand end, as in Figure 4.3a. Note that action potentials will only occur at the nodes, and that a finite time will elapse between the excitation of each node in the chain. Each pause represents the time (or electrotonic latency) needed to charge or discharge the membrane capacitors with the preceding nodal action potential current. This is called **saltatory**, or **stepwise**, conduction.

4.4 EXCITATION THRESHOLD

At any one time, an electrically excited fibre contains several conducting action potentials along a finite length, albeit at differing stages of evolution. If the speed of conduction of action potentials (θ cm sec^{-1}) is uniform and the duration of the action potential (λ sec) is the same at each site, the length of the activated region of the fibre will be $\theta\lambda$ cm. For safety of ongoing conduction the area of activated membrane must be of sufficient size to ensure that, at points of multiple branching, for example, downstream branches are electrotonically depolarized to threshold by the approaching action potential. Very small zones of excited membrane may fail to conduct to the branches because there is insufficient 'current-giving' capacity (I in Figure 4.4) to excite the large plasmalemmal area of downstream branches.

The rate of electrotonic depolarization varies with the strength of the applied current, as shown in Figure 4.1. A rather complex

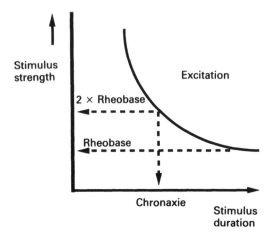

Figure 4.6 Minimum strengths and durations of depolarizing current pulses required to reach threshold voltage in a muscle fibre arranged in the manner shown in Figure 4.3a. Pulses of current with strength/duration co-ordinates above and to the left of the line shown will reach threshold and excite an action potential. Minimum current strength needed to reach threshold is called the **rheobase**. Current of twice rheobasic strength must flow continuously for a duration known as the **chronaxie** in order for the membrane to reach threshold.

relationship exists between the minimum current strength and the minimum current duration needed to reach threshold. This is shown graphically in Figure 4.6. A mathematical description of this relationship was devised by Noble and Stein (1966).

4.5 VARIATIONS IN THRESHOLD FOR EXCITATION

So far, the threshold voltage has been treated as a constant. Usually, however, electrotonic depolarization of the plasmalemma actually elevates threshold, as shown in Figure 4.7. Note that the original value of threshold was reached by the rising membrane potential during the depolarization step. Had the threshold voltage not risen also, an action potential would have occurred. As it was, the depolarization step in Figure 4.7 failed to excite an action potential because at all times the rising threshold voltage kept just above the level reached by the rising membrane potential. In Figure 4.8, however, the threshold is depicted as rising more slowly than in Figure 4.7, so that an action potential would occur at the arrow. Here the rising membrane potential briefly overtakes the more slowly rising threshold. Threshold voltage represents a point at which total

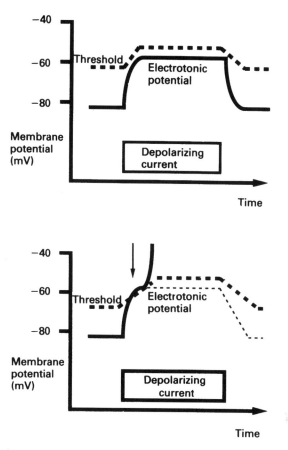

Figure 4.7 and 4.8 A muscle fibre is depolarized by the passage of a small transmembrane current during the period indicated, using the circuit shown in Figure 4.3a. During the time course of electrotonic depolarization there is a rise in threshold, which falls again when depolarizing current ceases. In Figure 4.7, at no point does electrotonic depolarization exceed threshold. In Figure 4.8, however, the threshold rises less steeply, and is exceeded by the membrane potential at the arrow, inducing an action potential.

membrane inward current exceeds total membrane outward current. The former can be considered to be carried mainly by I_{Na} and the latter by I_K. Depolarization will take the membrane potential closer to E_{Na} and further from E_K, as shown in Figure 3.3. The magnitude of the ionic current flowing through a particular ion channel of constant conductance is proportional to the difference between the prevailing membrane potential and the electrochemical diffusion potential for that ion, as discussed in connection with Figure 3.1. In this way, depolarization will increase outward I_K and reduce inward

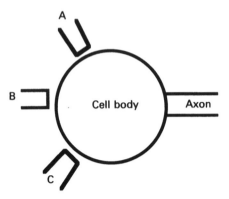

Figure 4.9 A nerve cell body with which dendritic processes from three other nerve cells (A–C) connect (or synapse).

I_{Na}, thereby raising the threshold. Depolarization tends to increase g_{Na}, however, (as discussed already in relation to Figure 3.3, and dealt with in more detail in Chapters 5 and 6) and this will lower the threshold. Threshold voltage is a resultant, therefore, of two opposing influences. If membrane potential and threshold voltage values diverge, the membrane becomes less easily excited, a phenomenon called **accommodation**. Alternatively, during a rapid depolarization, the membrane potential may become temporarily closer to threshold. The temporary increase in excitability so produced is called **latent addition**.

The mammalian nervous system has many points of functional contact, called synapses, between constituent nerve cells. Action potential pathways in the nervous system consist not only of simple one-to-one links between cells, but also of routes where many cells, each theoretically capable of carrying an action potential, impinge upon a single large nerve cell located in a final common pathway, as depicted in Figure 4.9. In this situation, in contrast to the former, an action potential in just one pre-synaptic nerve cell (**A–C**) may only slightly depolarize the adjacent large area of post-synaptic nerve cell. Threshold voltage may not be reached, so trans-synaptic conduction fails. In contrast, if several pre-synaptic nerve cells carry action potentials simultaneously, or nearly simultaneously, the resulting depolarization of the post-synaptic cell body may cause threshold to be reached and synaptic transmission to occur. This phenomenon is called **summation**, and it may operate either spatially or temporally. Synaptic transmission usually involves the release of an excitatory chemical from the pre-synaptic nerve terminal. The released chemical depolarizes the post-synaptic cell in a way similar to the type of electrotonic depolarization shown in Figures 4.1, 4.2, 4.3, 4.5, 4.7 and 4.8. The importance of this mode of cell-to-cell conduction of action potentials is that it can be blocked by pharmacological means.

4.6 REFRACTORINESS TO RE-EXCITATION

Following an action potential, the cell membrane becomes temporarily unresponsive to further electrical stimulation. Initially, the refractoriness is complete, with the cell unresponsive even to very large currents. The duration of this phase is called the **absolute refractory period**. Later, there is a period of partial refractoriness during which excitation may be achieved, but only by a larger current than was needed originally to elicit the action potential. An excitability threshold voltage now exists, but it is higher than usual. Later still, full excitability is restored. Immediately prior to this point, however, the cell may display a brief period of supra-normal excitability. Refractoriness represents a non-conductive state of the membrane Na^+ channels, and will be dealt with in more detail in Chapters 5 and 6.

A fully referenced account of the electrical phenomena that have been outlined in the present chapter can be found in the comprehensive reviews prepared by Carmeliet and Vereecke (1979) and by Fozzard (1979).

– 5

Regulation of plasmalemmal ion channel conductance

By definition, ion channels in electrically excitable membranes are opened and closed by transmembrane voltage changes. They are also regulated in a time-dependent manner. A combination of these two influences determines the time-course and voltage-course of an action potential. Analysis of their respective influences has been made possible by an experimental technique called **voltage clamping**. For this purpose, a pair of electrodes is arranged with one member of the pair on each side of the plasmalemma. Membrane potential is continuously sensed by the electrodes and is automatically kept at a chosen, constant value by means of the passage of requisite amounts of transmembrane current in the appropriate direction. At any chosen membrane potential, therefore, changes in transmembrane current can be recorded. Doubts have been expressed about the adequacy of membrane potential control that is attainable in this way in certain types of cells (Sommer and Johnson, 1968). Nevertheless, the technique can give acceptable membrane potential control when appropriate precautions are taken. Voltage control is particularly reliable when applied to isolated single cells. The magnitude of transmembrane clamping current needed to maintain the membrane potential at the chosen value is monitored continuously. This permits the time-course of ion currents to be studied before and after selected stepwise (or ramp) changes in clamp voltage.

5.1 INWARD AND OUTWARD TRANSMEMBRANE IONIC CURRENTS

Figure 5.1 shows the results of an experiment in which a piece of cardiac muscle was bathed initially (5.1a) with normal, Na^+-containing, extracellular fluid, and then later (5.1b) with Na^+-lacking fluid, sucrose being added to preserve isotonicity. The membrane was first clamped to a holding voltage of $-80\,mV$, near to

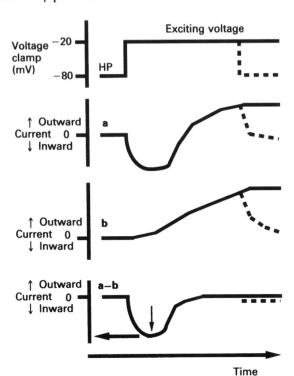

Figure 5.1 Inward and outward currents. A voltage-clamped muscle fibre was maintained at a holding potential (HP) of $-80\,mV$ and then step-depolarized to $-20\,mV$ (top panel). Transmembrane current, which is zero at the HP, becomes inward soon after stepping the voltage to $-20\,mV$ (line **a**), provided that the tissue is bathed with a solution containing Na^+. Note that transmembrane current becomes outward in direction as time progresses. When this experiment is repeated (line **b**) using an extracellular fluid which lacks Na^+, inward current is absent, although later outward current persisted. Inward Na^+-dependent current magnitude is shown in the bottom panel as the difference (**a−b**) between lines **a** and **b**. The dashed line at the right-hand end of each panel represents the change that occurs when the voltage clamp is stepped back to the HP.

the normal resting membrane potential for cardiac muscle. A subsequent stepwise depolarization to $-20\,mV$ induced a transient inward current, but only if the extracellular fluid contained Na^+, as shown by the calculated difference between current records **a** and **b**. The inward current that flowed is most readily explained by an influx of extracellular Na^+ across the plasmalemma, as discussed previously in connection with Figure 3.3. A spontaneous, time-dependent decay of this inward current occurred after the arrow in **a−b**, representing a phase of declining g_{Na} values. The remaining, persistent outward current is carried by K^+ (line b). In a voltage clamp circuit electric current flowing out of a cell, via the plasmalemma, (line b) requires an equal inflow of current via the intracellular electrode. As discussed in Chapter 1, however, unless the total number of anions changes in the intracellular compartment, any loss of cations from a cell, such as would be produced by transplasmalemmal efflux of K^+, must be matched by an equal influx of cations via the intracellular electrode. In fact, most intracellular glass microelectrodes are filled with a solution of KCl, so that electric

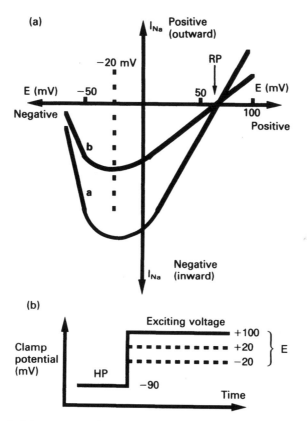

Figure 5.2a–b Maximum Na^+ currents. Current/voltage plots (5.2a) of a voltage-clamped muscle fibre during a series of step depolarizations from a fixed holding potential (HP) of $-90\,mV$, to various depolarized levels of membrane potential (E) indicated in 5.2b. Maximum Na^+ current (I_{Na}) was derived in each case from the arrowed point on the a–b line shown in the bottom panel of Figure 5.1. Depolarization to about $-20\,mV$ (line a, 5.2a) gives the largest inward I_{Na} value. Steps to E values above about $+60\,mV$ (the reversal potential, RP) generate outward I_{Na} values. An experiment conducted in the presence of a Na^+ channel blocking drug is shown as line b (5.2a)

current entering the cell via the intracellular electrode can be carried by K^+. In other words, there is a net transcellular K^+ flux. In a similar way, transplasmalemmal influx of Na^+ (line a) is associated with exit of cations from the cytoplasm into the tip of the microelectrode. Experiments may be performed in this way, using a fixed value of clamp voltage prior to the depolarization step, but varying the levels of membrane depolarization after the step. In Figure 5.2a the peak post-step I_{Na} values so produced are plotted against the corresponding post-step clamp voltages (E), using Na^+-containing extracellular fluid throughout and starting from a pre-step clamp voltage of -90 mV in each case. Knowing the post-step voltage (E) and the resulting peak I_{Na} values, one may calculate the peak g_{Na} values induced, using Ohm's law, as described in relation to Figure 3.1:

$$g_{Na} = \frac{I_{Na}}{E_{Na} - E}$$

where E_{Na} is the electrochemical diffusion potential for Na.

When $I_{Na} = 0$ it follows that $E_{Na} - E = 0$ and $E = E_{Na}$. Figure 5.2a shows that this applies at about $+65$ mV. In theory, this provides a method for calculating E_{Na}. In practice, however, this voltage should be termed the **reversal potential,** as it merely represents the transmembrane voltage above which inward current is replaced by outward current. Before concluding that this corresponds to E_{Na} it is necessary to demonstrate that the reversal potential is shifted along the voltage axis by appropriate manipulations of the transmembrane Na^+ concentration gradient.

Line b in Figure 5.2a represents an experiment where peak I_{Na} values were obtained in the presence of a local anaesthetic drug, which blocks Na^+ channels.

5.2 ACTIVATION CURVES

In Figure 5.3a the I_{Na} values from Figure 5.2 are replotted in the form of g_{Na} values (lines a and b). This plot, known as an **activation curve,** reveals that activation begins at about -70 mV and is maximal at about -40 mV. In some tissues activation begins even at normal resting potential. Local anaesthetic drugs displace the activation curve (line b) to less negative (more positive) voltages. In other words the tissue becomes less easily excited electrically, and the threshold rises. At higher concentrations most local anaesthetic drugs also reduce the maximum attainable g_{Na} values (line c). The cell then

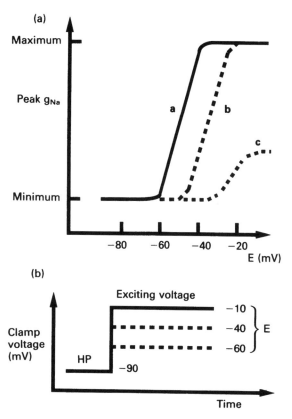

Figure 5.3a–b A re-plot of some of the data shown in Figure 5.2. Transmembrane voltages and associated g_{Na} values (5.3a) were derived from a voltage-clamped muscle fibre during a series of step depolarizations from a fixed holding potential (HP) of -90 mV, to the various depolarized levels of membrane potential (exciting voltage, E) indicated in 5.3b. As discussed in connection with Figure 3.3, Ohm's law may be used to calculate g_{Na} (line **a**) from the I_{Na} values (inward negative) shown in Figure 5.2, since $g_{Na} = I_{Na}/(E - E_{Na})$, where E_{Na} is taken as the reversal potential ($+60$ mV). Lines **b** and **c** were from similar experiments conducted in the presence of a Na^+ channel blocking drug at a low and a high concentration respectively.

becomes not only less easily excited but also less responsive once it has been excited.

5.3 INACTIVATION CURVES

Voltage clamping readily permits the voltage-dependence of ion channel closure to be studied. To do this, the membrane needs to be clamped in turn to a series of different conditioning or holding voltages, representing various levels of depolarization relative to the

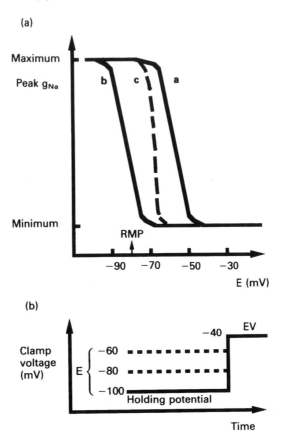

Figure 5.4a–b Inactivation of g_{Na}. Transmembrane voltage (E) and associated peak g_{Na} values (5.4a) generated by a voltage-clamped cardiac muscle fibre following step depolarization to a fixed exciting voltage (EV) of -40 mV from various holding potentials, as indicated in the protocol of 5.4b. Values of peak I_{Na} were derived as shown at the arrow in the bottom panel of Figure 5.1. Peak I_{Na} values were converted to peak g_{Na} values using Ohm's law, as indicated in the legend to Figure 5.3. A typical resting membrane potential (RMP) value is shown. Inactivation of g_{Na} in line **a** begins at voltages positive to RMP. A similar experiment conducted in the presence of a Na^+ channel blocking drug (line **b**) shows inactivation beginning at potentials somewhat negative to RMP. With a different Na^+ channel blocking drug (line **c**), inactivation of peak g_{Na} does not occur at the RMP but is potentiated at more positive potentials (e.g. between -65 and -70 mV).

normal resting membrane potential. The membrane is then stepped each time to a single level of depolarization (e.g. -40 mV, Figure 5.3a), chosen because it is known to raise g_{Na} maximally. A suitable clamp protocol is shown in Figure 5.4b. The peak g_{Na} values attained after the depolarizing step (to -40 mV) depend upon the preceding holding or conditioning voltages, as shown in Figure 5.4a. Holding the membrane at conditioning voltages positive to normal resting potential (i.e. positive to about $-80\,mV$) prior to this step usually reduces the maximal g_{Na} value that is attainable after the step. This plot, known as an **inactivation curve**, shows that inactivation is more or less complete after clamping the membrane to a value of $-50\,mV$ or above, with a state of permanent refractoriness having been created. Line **b** in Figure 5.4 shows that a local anaesthetic drug displaces the inactivation curve to more negative potentials, so that inactivation may begin even at the normal resting membrane potential (RMP) value. A raised $[Ca^{2+}]_o$ usually has the opposite effect. That is to say, it renders Na^+ channels less sensitive to

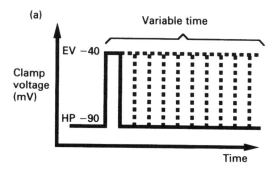

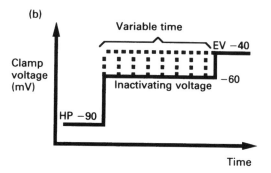

Figure 5.5a–b Voltage clamp protocols for investigating the time-dependence of g_{Na} activation (5.5a) and inactivation (5.5b). In both cases the cell is maintained at a holding potential (HP) of $-90\,mV$ prior to step depolarization. In 5.5a, the step depolarization is to the g_{Na} exciting voltage (EV) of $-40\,mV$ (Figure 5.3) for between 0.1 and 1.0 ms. In 5.5b step depolarization was first to the g_{Na} **inactivating voltage** of $-60\,mV$ (Figure 5.4) for between 1 and 10 ms, followed by a second depolarization step to $-40\,mV$. Maximum g_{Na} after the second step provided a measure of the extent to which Na^+ channels have been inactivated by the preceding time spent at $-60\,mV$.

depolarization-induced inactivation (Dumaine *et al.*, 1990). Alternatively, some drugs of the local anaesthetic type, particularly at low concentrations, may fail to alter the inactivation threshold or the conductance of Na^+ channels held at a normal resting membrane potential, but will potentiate inactivation at a moderate level of membrane depolarization, as shown by line c. Since injured tissues are usually depolarized, this latter property confers upon such drugs a degree of selectivity towards injured cells, as discussed further in Chapter 9. Note also that because $[K^+]_o$ determines resting membrane potential (E) in equation 1.9, drugs showing selectivity towards depolarized cells will have pharmacological actions which depend upon the level of serum $[K^+]$. The therapeutic implications of this fact will also be discussed in Chapter 9.

Theoretically, the time-dependence of opening and closing events in ion channels should be analysable by means of voltage clamping. Clamp protocols for this purpose are shown in Figure 5.5. Early currents created after a voltage clamp step, however, usually contain large capacitative currents. This makes it difficult to separate out any co-existent early resistive components. The time constant for inactivation is usually sufficiently long to be analysed in this way, but that for activation is too brief in the case of most Na^+ channels at physiological temperatures.

5.4 WINDOW CURRENTS

Note that lines **a** (representing g_{Na}) in Figures 5.3 and 5.4 would have intersected their lower ends at a membrane potential of about $-65\,mV$ if both had been plotted on the same graph. Prolonged clamps to voltages between $-60\,mV$ and $-70\,mV$, therefore, cause rapid partial activation of g_{Na} and incomplete subsequent inactivation. Inward I_{Na} continues indefinitely at these voltages, causing a so-called **window current**. Drugs which block Na^+ channels tend to move the g_{Na} line rightwards in Figure 5.3 and leftwards in Figure 5.4. It follows that the voltage range within which the lower ends of the two g_{Na} lines 'overlap', would narrow and finally vanish (lines **b**) under the influence of these drugs. Such drugs therefore block Na^+ window currents (Attwell *et al.*, 1979). This may be responsible for their ability to promote repolarization and hence abbreviate the cardiac action potential.

– 6

Models and mechanisms of ion channels

Several attempts have been made to express the opening and closing kinetics of ion channels in mathematical terms. Hodgkin and Huxley (1952) provided a set of empirical equations that closely fitted the observed behaviour of Na^+ channels in a nerve. Although these provided few insights into the actual mechanisms involved, it is significant that activation of conductance in their equations had to be expressed by a dimensionless variable m, whereas inactivation of conductance was expressed by a quite different dimensionless variable h. Both variables represented the probability of appropriate parts of the conductance machinery being present in the channel at a given moment. The concept that inactivation is a separate process from activation, and not merely a reversal of activation, was novel and is still important to our understanding of ion channels.

6.1 GATING MECHANISMS

Haas *et al.* (1971) went further, by suggesting that Na^+ channels contain, and could be blocked by, moveable gates, of which two types exist in each channel. These gates are structural analogues of the m and h variables in Hodgkin and Huxley's equations, and are so labelled in Figure 6.1. The position of the gates determines the conductance of the ion channel. Three more or less distinct states may be envisaged, often termed **resting, open** and **inactive** (Figure 6.1). The latter two are considered to be **non-resting**, and only the open form has a high conductance. Each gate may remain in a fixed position (open or closed), or it may alternate between open and closed positions. Transition between open and closed positions seems to be instantaneous, but different periods of time are spent open after successive openings, just as different periods of time are spent closed after successive closings. One can measure the average open time (or closed time) for a particular gate, and the statistical probability that a particular gate will be open (or closed) at any given moment. The open position of a particular m gate is usually maintained only briefly, although pharmacological agents are available which prolong

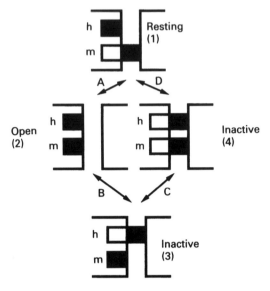

Figure 6.1 Gating of Na$^+$ channels. Four 'states' in a plasmalemmal Na$^+$ channel, having an activation gate (m) and an inactivation gate (h). At any given moment a particular gate is either fully open or fully closed, but may alternate between open and closed positions. In a normally polarized resting cell membrane, however, m gates remain closed and h gates remain open, with neither gate alternating position (state 1). When both gates in a channel are open (state 2) there is high g_{Na}, but when either gate is closed (states 1, 3 and 4) there is low g_{Na}. Under voltage/time conditions which selectively increase g_{Na} in a previously resting channel (Figure 5.3), the m gate begins to alternate position at high frequency, but with a short mean open time, thereby increasing open probability of the channel (arrow A). In contrast, under voltage/time conditions that selectively inactivate g_{Na} in a previously resting channel (Figure 5.4), the h gate begins to alternate position. This occurs at low frequency but with a prolonged mean closed time (arrow D). The difference between states 1 and 4 is that in the latter state any conversion to state 2, and thus any increase in g_{Na}, requires opening of both m and h gates. Voltage/time conditions which induce h gate closure, therefore, need to be reversed before m gate opening increases time-averaged g_{Na}. Meanwhile, the tissue will remain refractory to electrical stimulation. During the early stages of an action potential, Na$^+$ channels develop increased open channel probability as a result of alternations between states 1 and 2 via arrow A. Because of the consequent membrane depolarization, however, there is time-dependent h gate closure, due to alternation between states 2 and 3 via arrow B, with an attendant decrease in the open probability of the channel, and development of a refractory period. Restoration of excitability (Figure 6.2) requires time- and voltage-dependent re-opening of h gates.

this considerably. Other drugs leave open time unchanged but increase open state probability by causing more opening events per unit of time. The conductance of a single open channel is termed its **unitary conductance**. Unitary conductance multiplied by the open state probability determines the observed time-averaged conductance. Both voltage- and time-dependence of m and h gate alternation may be manipulated with drugs. These are responsible for depolarizing and hyperpolarizing shifts shown by line b in Figures 5.3 and 5.4 respectively.

The h gates are sluggish by comparison with the m gates. That is to say, they have longer open and closed times than the m gates when both gates are alternating position. In a basal or resting state the h gates of Na$^+$ channels seem to be permanently open and the m gates permanently closed. Activation of Na$^+$ channel conductance from a resting to an open state (state 1 to state 2, arrow A) causes the m gates to alternate, that is to say, they open frequently but usually briefly. Inactivation, or movement of the h gates from an open to a closed position, takes place abruptly, but when it has occurred the gates remain closed for a considerable time. Likewise, an alternating h gate spends a considerable time open before it closes again. When the h gate is closed the channel remains in a low conductance or inactivated state, irrespective of the position of the m gate (states 3 or 4).

6.2 PATCH-CLAMPING

Evidence for the stochastic type of ion channel gating behaviour described above has been derived using a technique called **patch-clamping** (Scanley *et al.*, 1990). For this purpose a hollow glass tube is drawn out to a fine hollow tip and filled with electrolyte solution. Across the tip of this tube a tiny patch of a suitable cell membrane may be sealed. The membrane area may be made small enough to contain, in some cases, just a single ion channel. Application of a transmembrane voltage which is known from classical voltage-clamping studies to activate g_{Na} (Figure 5.3), causes single channel conductance recorded in a patch-clamp experiment to begin to alternate at high frequency between fixed high and low values. Presumably the m gate is alternating between states 1 and 2 in Figure 6.1. Conductance changes continue to be observed as long as the h gate remains open. Under conditions where the h gate also begins to alternate (i.e. begins to inactivate conductance), assuming that the m gate continues to alternate rapidly, periods occur when measured single channel conductance continues to alternate rapidly between high and low values (arrow A). Other periods will occur intermittently when uniformly low conductance prevails (states 3 and 4, arrow C). Any given cell contains many Na^+ channels, so that whole cell g_{Na} values measured in a classical voltage-clamp, for example, will be graded, depending upon the time-averaged conductance of the entire unsynchronized population of channels.

The gates on ion channels that determine conductance probably represent specific parts of protein molecules that line the channels. Particular proteins that are integral to Na^+ channels have been identified in several electrically excitable cell membranes. Ion-channel proteins from different organs, and even different vertebrate species, display considerable structural resemblances to each other (Lombert and Lazdunski, 1984). These proteins would seem to provide the basis for Na^+ channel conductance because, when they are incorporated into artificial lipid membranes, they confer many of the conductance properties that were seen in the cells from which the proteins were derived. Specific parts of these channel proteins confer voltage sensitivity, other parts confer ion selectivity, while yet other parts are responsible for the gating properties. Recent progress in this area has been reviewed in a book edited by Narahashi (1990). The gene which controls the synthesis of one of these channel proteins has been sequenced by Noda *et al.* (1984), and it has even been cloned. Monoclonal antibodies are now available against the protein, which should prove useful to investigators of Na^+ channels.

6.3 THE MODULATED-RECEPTOR HYPOTHESIS

Hondeghem and Katzung (1977 and 1984) proposed that the binding affinity of local anaesthetic drug molecules to Na^+ channels depends upon the open or closed position of channel gates, and thus on the prevailing conductance state. Note that cause and effect operate in both directions. The extent of drug binding depends upon the position of the channel gates, but in turn it also influences the gate positions. This is termed the **modulated-receptor hypothesis**. The hypothesis was formulated primarily to account for observed variations in the ability of certain local anaesthetic drugs to block Na^+ channels, depending upon whether the cell membrane was quiescent or had been generating action potentials prior to the test stimulus. The first action potential after a period of electrical quiescence is often more or less unaltered by the presence of a local

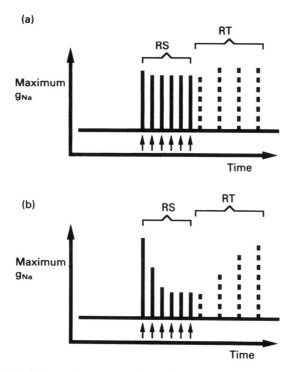

Figure 6.2a−b Use-dependent g_{Na}. Decline of maximum g_{Na} values (vertical lines) during a train of six action potentials induced in a myocardial fibre by six repetitive electrical stimuli (RS, at the arrows) in a previously quiescent cell. Figure 6.2a was from an experiment performed in the absence of a drug, whereas 6.2b was recorded after exposure of the muscle to a low concentration of a Na^+ channel blocking drug. The decline in maximum g_{Na} values with successive action potentials in the train is less prominent in 6.2a than in 6.2b. Following the sixth stimulus in the train there is a pause of variable duration before a final (seventh) stimulus is applied. The duration of this pause constitutes the recovery time (RT). The maximum g_{Na} value attained during the seventh action potential is plotted as a vertical dashed line.

anaesthetic drug. Subsequent action potentials in a train, as depicted in Figure 6.2, may become progressively more blocked. The local anaesthetic potentiates a normal 'fading' phenomenon (Figure 6.2b) that was only just detectable in responses of a non-drug treated cell (Figure 6.2a). Drug-induced potentiation of fading becomes more noticeable the higher the frequency of stimulation. If drug molecules possess a high binding affinity for Na^+ channels only when the latter are in a non-resting state (Figure 6.1), a previously quiescent membrane would be expected to respond normally to the first action potential thereafter, because drug molecules would have become dissociated from resting Na^+ channels during the preceding period of quiescence. The non-resting channels created by the first action potential, however, would permit more drug binding to occur, with progressive block of subsequent action potentials in the train (Figure 6.2b). This phenomenon is termed **use-dependent block**. Recovery from use-dependent block is portrayed by the group of vertical dashed lines in the figure. The g_{Na} response line at the extreme right-hand side of Figure 6.2b was elicited by a single stimulus delivered after use-dependent block had nearly disappeared. The other g_{Na} responses, shown by the series of vertical dashed lines in this figure, were each elicited by a single stimulus delivered sometime earlier during recovery. From these responses the kinetics of recovery may be studied. It is usually a mono-exponential or bi-exponential process. The former may represent preferential binding of local anaesthetic drug molecules to just one of the non-resting states of the Na^+ channel. This situation provides an opportunity to selectively prolong the refractory period of the myocardium, for example, and to prevent one action potential occurring in dangerously close proximity to another, as will be discussed further in Chapter 9. Most local anaesthetic drugs, however, also cause some of the other changes shown in Figures 5.2–5.4, so that their selectivity is only partial. Recovery of excitability after an action potential represents the conversion of the non-resting Na^+ channels in Figure 6.1 into resting ones.

The myocardial absolute refractory period usually ends prior to the end of the action potential which has caused it. Some repolarization is needed, however, before recovery of excitability begins. The kinetics of g_{Na} reactivation in mammalian myocardium, as depicted in Figure 6.2, were studied in great detail by Gettes and Reuter (1974). Contrary to the situation in nerves, they found that reactivation is a substantially slower process than the process of inactivation of g_{Na}. It is accelerated by a rise in $[Ca^{2+}]_i$, which has therapeutic significance, since it means that the higher the level of $[Ca^{2+}]_i$ in a cell the shorter will be its refractory period. This has considerable bearing upon certain cardiac rhythm disturbances and their pharmacological treatment (Chapter 9).

In the presence of a local anaesthetic drug of the type shown in Figure 6.2, the refractory period may considerably outlast the action potential. The drug appears to exert an opposite effect to a rise in $[Ca^{2+}]_i$. Reopening of the h gates of Na^+ channels at the end of an action potential is rendered less probable by the local anaesthetic at any given point in time, and after a given set of preceding voltage conditions. This may be due to a briefer h gate open time or to a lower h gate open probability due to lower frequency of opening. Both mechanisms appear to contribute to delayed recovery of excitability in the case of most drugs of this type.

– 7

Ion currents other than the sodium current that participate in action potentials

In most electrically excitable cells conductance changes occur in several distinct types of ion channel during each action potential. They have been explored with the aid of voltage clamps in ways analogous to those shown in Figure 5.1 for Na^+ channels. Conditions are chosen which, it is hoped, will selectively activate or block just one type of channel at a time. Alternatively, attempts are made to block all but one of the existing channel types. It is never possible to be sure, of course, that either objective has been achieved, and many cross-checks are necessary. Selective inactivation of one type of channel by clamping the cell to a particular conditioning or holding voltage, as in Figure 5.4, is a useful stratagem. Allowing a cell to remain at a particular clamp voltage for a long period sometimes ensures that a particular time-dependent conductance change becomes fully activated, and that others become fully inactivated. In Figure 5.1, line a, for example, I_{Na} was first activated, but had become fully inactivated at the right-hand side of the record, leaving an uncontaminated outward current. The latter also had been time-dependently activated, but conductance had persisted. When the voltage clamp is stepped back to its original holding potential value, somewhat positive to E_K, the outward current time-dependently offsets, creating a so-called current 'tail', shown by the dashed lines in Figure 5.1 (lines a and b).

Drugs with a selective blocking or stimulant action upon a particular channel are valuable tools for dissecting a mixture of ion currents (Figure 5.3). Unfortunately, their selectivity is often hard to prove, and frequently is less complete than one would wish. Removal of a particular ion from the extracellular fluid is an attractively simple way to prevent an inward current being carried by that ion (Figure 5.1). It is sometimes possible to achieve a similar manipulation of the composition of the intracellular fluid. After several decades of experiments by many workers, a composite and widely accepted picture of the principal ion conductance changes during an action potential has emerged. This is summarized in Figure 7.1 for a typical nerve cell and a typical mammalian cardiac muscle cell.

7.1 COMPARISON OF ACTION POTENTIALS IN DIFFERENT TISSUES

Marked differences occur in the time course of action potentials in different tissues. Those in nerves are briefer as a rule than those in the myocardium (Figure 7.1). Termination of cardiac action potentials, or **repolarization** as it is called, is delayed partly by a slow rise in g_{Ca}, but then promoted by a subsequent fall in g_{Ca}. The Ca^{2+} channels involved are termed slow Ca^{2+} channels for this reason. The slow and delayed secondary rise in g_{Na} that is seen during a cardiac action potential (Figure 7.1c) is usually attributed to the movement of Na^+ through slow Ca^{2+} channels, suggesting that these channels do not discriminate between the two ions. The experimental conditions needed to detect such Na^+ movement, however, may have reduced a normally existing ion selectivity (Campbell and Giles, 1990). More than one type of Ca^{2+}-conducting channel probably exists. Not all types display the slowness of activation and inactivation that is depicted in Figure 7.1c. The faster and more transiently opening channels are given the name T channels, to distinguish them from the classical, longer acting L channels.

Just as fast Na^+ channels contain moveable gates, so also do some of the other channels depicted in Figure 7.1. The activation and inactivation gates that have been identified in slow Ca^{2+} channels, for example, are designated d and f gates respectively. The membrane voltages required to induce gate movement in slow Ca^{2+} channels, however, are less negative (or less polarized) than those needed to operate equivalent gates in fast Na^+ channels. Action potentials in the atrioventricular node of the heart, and in most types of mammalian smooth muscle, are particularly strongly dependent upon slow Ca^{2+} channels for the carriage of inward current. Drugs like diltiazem selectively block the slow Ca^{2+} channels, and such drugs tend to relax smooth muscle and interfere with conduction of action potentials through the atrioventricular node of the heart. In ways closely analogous to those shown in Figures 5.2–5.4, however, myocardial slow Ca^{2+} channels may also become selectively blocked by such drugs, producing either a rise in threshold voltage or a

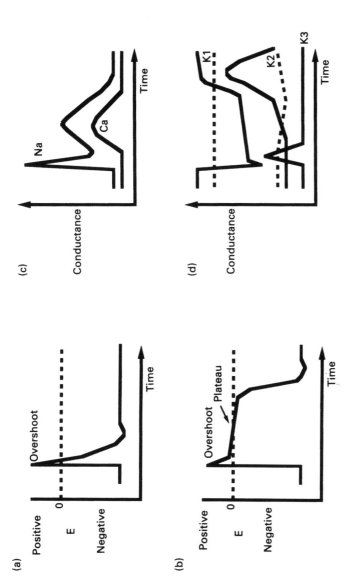

Figure 7.1a–d Dissection of mixed ionic conductances. Time course of changes in membrane potential (E) and ionic conductance (g) during an action potential in a nerve (7.1a) and in a myocardial cell (7.1b, c and d). Action potential regions positive to zero voltage are called **overshoots**. An action potential in a myocardial cell, although not in a nerve, displays a prominent plateau phase, during which E approximates to zero. Changes in g_{Na} and g_{Ca} during a myocardial action potential are shown in 7.1c, with concomitant g_K changes shown in 7.1d. Three K⁺ channel conductances are recognized, labelled here g_{K1}, g_{K2} and g_{K3} respectively. The simplest (g_{K3}) is termed the **transient outward K⁺ conductance**. The other two display rectification, their conductance depending upon the direction of ionic current. The first (g_{K1}) is called the **immediate inward rectifier**. The second (g_{K2}) is known as the **delayed outward rectifier**. Outward g_K values in 7.1d are shown as continuous lines and inward g_K values as discontinuous lines.

prolongation of the refractory period. Prominent ion-dependent and voltage-dependent block is displayed by some members of this group of drugs. Blockade of conduction through the atrioventricular node by drugs of this type may be of therapeutic use in some situations, but can also be an unwanted side effect in other contexts.

7.2 SODIUM–CALCIUM EXCHANGE PROCESSES

Most myocardial cells display a Na^+ for Ca^{2+} exchange across the plasmalemma. Similar exchanges have been found in many other tissues. The stoichiometry usually involves three Na ions for each Ca ion moved, so net current flows in the direction in which Na^+ moves and the process is said to be electrogenic. Exchanges can operate in either direction across the plasmalemma depending upon the prevailing membrane potential and ion concentration gradients. Gating properties have not been discovered. A rise in $[Na^+]_i$ will promote Ca^{2+} entry, whereas a rise in $[Ca^{2+}]_i$ will promote Na^+ entry. The former occurs, for example, when myocardial $[Na^+]_i$ rises during Na pump inhibition with the drug digoxin. This also explains why digoxin enhances intracellular stores of Ca^{2+}, and hence the force of myocardial contraction. On the other hand, entry of Na^+ and loss of Ca^{2+} occur when there is a rise in $[Ca^{2+}]_i$. This occurs in the myocardium during each systole, mainly because Ca^{2+} is released into the cytoplasm from intracellular stores. Removal of Ca^{2+} from the cell, by exchange with extracellular Na^+, helps to balance the entry of Ca^{2+} that occurs via slow Ca^{2+} channels during an action potential (Figure 7.1). This exchange generates an inward current, however, which in turn delays repolarization, or may even generate a secondary depolarization after the cell has already been fully repolarized. This is called an **after depolarization**, and can induce another action potential if it reaches threshold. It is an important cause of disturbances of cardiac rhythm during states of Ca^{2+} overload (Chapter 9).

7.3 RECTIFICATION IN ION CHANNELS

In most tissues any enforced depolarization of the plasmalemma induces K^+ efflux from the cell, for the reasons discussed in Chapter 3. Some K^+ channels, however, display a conductance that depends upon the direction of current flow. This phenomenon is called **rectification** and is illustrated in Figure 7.1d. The equivalent electrical

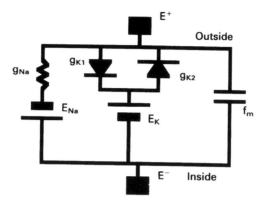

Figure 7.2 Rectification. An equivalent electrical circuit for a myocardial cell plasmalemma containing K^+ channels of two types, with conductances of g_{K1} and g_{K2} respectively. The former rectify inwardly and the latter outwardly, as depicted by rectifier symbols of opposing polarity. This circuit is a modification of Figure 3.1. Membrane potential (E) is a resultant of electrochemical diffusion potentials for Na^+ (E_{Na}) and for K^+ (E_K), operating through their respective conductances (g). Membrane capacitance is shown as f_m.

circuit is shown in Figure 7.2. Inward rectification indicates that when the rectifier is operational the inward conductance is high and the outward conductance is lower. Outward rectification signifies that when the rectifier is operational the outward conductance is high and the inward conductance is lower. Following a depolarization of the plasmalemma in the myocardium there is an extremely rapid decline in outward g_K (Weidmann, 1955; Hutter and Noble, 1960). On the other hand, following both depolarization and hyper-polarization of the plasmalemma inward g_K remains unchanged. These changes in g_K are said to represent inward rectification. Because rectification is induced so rapidly by depolarization, the rectifier is said to be an **instantaneous** or **immediate rectifier**, and is thought to operate non-time dependently in the sense that once it is made operational by depolarization the low outward g_K persists for as long as depolarization remains. At normal resting membrane potential, however, the rectifier is considered to be non-operational, so that high g_K is available in both inward and outward directions. It is the movement of K^+ through this type of channel which provides the normally high resting g_K of myocardial cells, sometimes spoken of as the **background conductance**. Measurement of inward and outward g_K values at various transmembrane voltages is derived from the sort of experiment shown in Figure 7.3. Reduction of

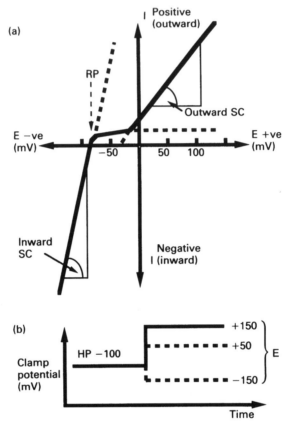

Figure 7.3a–b Rectification. Membrane potentials and maximum current values (I) recorded during prolonged post-step voltage clamps in a cardiac Purkinje fibre bathed with a Na^+-free solution. The fibre was clamped at a holding potential (HP) of -100 mV before being stepped (up or down) to various membrane potentials (E) between -150 mV and $+150$ mV (7.3b). This generates an I value equivalent to that at the right-hand end of line **b** in Figure 5.1. At values of E negative to the reversal potential (RP), which has a value of about -90 mV, the **inward slope conductance** (SC) shows a substantial gradient ($\Delta I/\Delta E$, or the reciprocal of slope resistance, from Ohm's law). At values of E positive to RP, however, slope conductance vanishes due to inward rectification of K^+ channels. At values of E above about -30 mV an outward SC appears, due to opening of K^+ channels.

outward g_K is probably due to plugging of the channel with Mg^{2+} from the cytoplasmic side.

The wide prevalence of inward rectifier K^+ channels in mammalian cells suggests that they are important, probably by limiting losses of K^+ that would otherwise occur during depolarization, as happens during each action potential. They assume particular importance in the heart due to the prolonged plateau stage of the action potential, during which the cell remains at a voltage markedly positive to the E_K value, and during which, therefore, K^+ efflux would otherwise occur. In order to maintain K^+ equilibrium in the long term, the cell must pump K^+ inwards at the expense of adenosine triphosphate to keep pace with diffusional losses. Economies in this direction, therefore, are probably advantageous, particularly when a cell is nutritionally deprived and adenosine triphosphate is scarce, leading to depolarization of the cell. This form of rectification, with its low outward g_K during depolarization, also serves to enhance depolarization induced primarily by a rise in g_{Na} during the early stage of an action potential, for reasons discussed previously in connection with equation 3.2.

Recent studies using single myocardial cells have required some revision of the ideas about inwardly rectifying K^+ channels that were described above. A number of workers have detected time-dependent onsets of inward rectification following depolarization of the plasmalemma. Onset was usually very rapid, but certainly not instantaneous (Sackmann and Trube, 1984a and 1984b; Kurachi, 1986; Shibasaki, 1987; Harvey and Ten Eick, 1988; Kilborn and Fedida, 1990; Sanguinetti and Jurkiewicz, 1990). In some cases the high inward g_K following a hyperpolarization step was not well maintained. That is to say, it time-dependently waned, in contrast to the behaviour expected of classical inward rectifier channels. Various names have been coined to describe these unusual forms, but at the present time it seems preferable to merely refer to the group as the **time-dependent inward rectifiers**.

7.4 ACTION POTENTIAL REPOLARIZATION

Action potential repolarization in nerves is completed soon after the peak of depolarization has passed (Figure 7.1a). In the myocardium, however, repolarization is often a two-stage process (Figure 7.1b). Early partial repolarization, prior to the plateau, is dependent upon two events, namely a rapidly time-dependent offset in high g_{Na} and a rapidly time-dependent onset of high g_K. Note that offset in g_{Na} may be incomplete, however, due to window currents (Chapter 5). It is conventional to consider the K^+ channels through which this early outward K^+ current flows to be of a distinct type, termed **transient outward** K^+ channels (Figure 7.1d). They are activated by the early reversal of membrane voltage during the upstroke of an action potential, from an inside negative to an inside positive value. This constitutes the 'overshoot' region in Figures 7.1a and 7.1b. The early outward I_K rapidly offsets in a time-dependent manner in most myocardial cells, leaving the membrane still substantially depolarized. This ushers in the plateau phase of the cardiac action potential. In cells having little plateau phase to their action potentials, such as those in most adult mammalian atria or in rat ventricles (Kilborn and Fedida, 1990), transient outward I_K is responsible for more of the overall process of repolarization. It is unlikely, however, that this current is important in cells with a well-developed plateau. Indeed, it may be totally absent from some myocardial cells (Clark *et al.*, 1988; Giles and Imaizumi, 1988; Nakayama and Fozzard, 1988; Hiraoka and Kawano, 1986, 1989). These channels may, however, contribute to action potential shortening during tachycardia. This aspect is dealt with further in Chapter 8.

Repolarization at the end of the plateau phase of most cardiac ventricular action potentials is required for the restoration of electrical excitability. The ion channels responsible for this repolarization are distinct from those responsible for the earlier, pre-plateau phase of repolarization. During the late phase of a ventricular action potential there is a gradual decline in g_{Na} and g_{Ca} (Figure 7.1c). This represents time-dependent offset of slow Ca^{2+} channel conductance, which causes repolarization for the reasons that have been discussed previously in connection with equation 3.2. Of even greater importance, however, is that there is also a time-dependent rather slow and delayed rise in outward g_K (Figures 7.1d, 7.2 and 7.3), termed **delayed outward rectification**. The K^+ channels concerned probably have only modest g_K values in both inward and outward directions at a normal resting membrane potential. Onset of high outward g_K without change in the low inward g_K appears in this sort of K^+ channel in a time-dependent manner at membrane potentials corresponding to, or above, plateau values of the cardiac action potential (Figure 7.3). Thus it is the plateau stage voltages that stimulate the final phase of action potential repolarization. Once activated in this way, however, high outward g_K is classically thought to persist without waning for as long as depolarization continues, as it would, for example in a voltage clamp. In contrast, under more physiological conditions, the repolarization that results from the rise in outward g_K would time-dependently close this ion channel in early diastole. This is an important contributor to pacemaker potentials, and will be dealt with in detail in Chapter 8.

Slow declines in outward I_K that have sometimes been observed by experimenters using prolonged clamps at depolarized voltage in multicellular tissues have traditionally been ascribed to accumulation of K^+ in the confined extracellular spaces just outside the plasmalemma. Isolated single cell studies can largely avoid this problem. These suggest that a genuine time-dependent offset in outward g_K in this sort of channel can occur, albeit slowly, under certain voltage clamp conditions. This means that if repolarization fails to occur for some reason, the period of high outward g_K will be only temporary. The cell may be able, therefore, to conserve its stores of K. Some investigators attribute time-dependent offset properties of g_K during prolonged periods of depolarization to events in a unique type of K^+ channel. Channels with and without such time-dependent offsets in conductance, however, are not sufficiently dissimilar to justify giving them separate names at this stage. Furthermore, K^+ channels with high outward g_K within the plateau range of voltages, which show conductance onset and offset kinetics intermediate between those of classical delayed rectifier and so-called transient outward types, have now been described by Yue and Marban (1988).

7.5 PHARMACOLOGY OF K⁺ CHANNELS

Barium and tetraethylammonium ions block the various K^+ channels present in the myocardium rather non-selectively (Hirano and Hiraoka, 1986). A number of other agents fairly selectively inhibit outward g_K, however, in the delayed outwardly rectifying K^+ channels during the plateau stage of a cardiac action potential, thereby postponing repolarization and prolonging the action potential. They may be used therapeutically to correct certain disturbances of cardiac rhythm (Chapter 9). The electrical and pharmacological properties of K^+ channel-acting drugs have been reviewed recently by Nakayama and Fozzard (1988), Berger *et al.* (1989), Dukes and Morad (1989), Colatsky *et al.* (1990), Kass *et al.* (1990), McCullough *et al.* (1990) and Sanguinetti and Jurkiewicz (1990).

In some mammalian tissues g_K depends upon the intracellular concentration of adenosine triphosphate. Very low concentrations of this substance, which may occur during periods of hypoxia or ischaemia, greatly increase g_K. A distinct type of K^+ channel, which shows little voltage sensitivity, is probably involved. Changes in inward current-carrying conductances at the start of the action potential will then produce less depolarization than previously (equation 3.2). Threshold voltage will rise further above the resting membrane potential and electrical excitability is reduced, for the reasons discussed in connection with Figure 4.7. Myocardial action potentials, in particular, become abbreviated due to the earlier start of repolarization. Equation 3.2 shows that repolarization will occur as soon as membrane potential becomes dominated by the outward current-carrying conductance channels, as it is in diastole. The higher g_K becomes, the earlier in an action potential will this point occur. Smooth muscles, in particular, become hyperpolarized. The physiological significance of these channels remains unclear. Drugs such as pinacidil, however, selectively activate conductance in such K^+ channels, thereby relaxing smooth muscle. This can be of therapeutic use in the treatment of arterial hypertension. Unfortunately, K^+ channel-opening drugs also have been shown to have a mixture of proarrhythmic and antiarrhythmic effects in laboratory animal hearts (Grover *et al.*, 1990). Other drugs, such as glibenclamide, inhibit opening of these K^+ channels during periods of adenosine triphosphate depletion. To what extent such inhibitor drugs protect the heart against the harmful consequences of adenosine triphosphate depletion is not yet established. If the role of these channels is primarily protective by shortening action potentials during periods when adenosine triphosphate is scarce, the inhibitors may even be deleterious (Wilde *et al.*, 1990). Shortening the duration of an action potential reduces the quantity of Na^+ and Ca^{2+} that subsequently must be pumped across the plasmalemma at the expense of already depleted adenosine triphosphate stores (Chapter 2).

Some K^+ channels possess a conductance that is increased by a rise in $[Ca^{2+}]_i$ (Tohse, 1990). Various investigators have given special names to these channels, but as in most respects they resemble the classical types that were described above, separate names are not presently justified.

A fully referenced account of the early literature of cardiac depolarization and repolarization research is contained in the review by Scher and Spach (1979).

– 8

Spontaneous action potential generation, with special reference to the heart

After the passage of an action potential the membrane potential in most electrically excitable cells settles down to a stable resting value. Those tissues which spontaneously generate action potentials, however, display a gradual depolarization at rest, until threshold is reached once again, and another action potential occurs. This spontaneous depolarization is termed a **pacemaker potential**. In mammalian hearts the phenomenon is particularly rapid in the normal pacemaker region called the sinoatrial node. It occurs also, albeit more sluggishly, in the atrioventricular node, bundle of His and Purkinje fibres. Many mildly injured myocardial fibres also display this phenomenon, and it is responsible for some common disturbances of cardiac rhythm, as discussed in Chapter 9. The pacemaker with the highest frequency of action potential production normally supplies the whole heart with a succession of action potentials at regular intervals. This also serves to co-ordinate pumping activity in the various cardiac chambers. Atrial systole occurs just ahead of ventricular systole because the action potential spreads out from the sinoatrial node through the atria, thence to the atrioventricular node and bundle of His, finally reaching the ventricles via the Purkinje fibre branches of the bundle of His.

8.1 IONIC CONDUCTANCE IN PACEMAKER POTENTIALS

For reasons discussed previously in connection with Figure 3.1 and equation 3.2, any time-dependent increase in conductance of an ion channel carrying current inwards, or any time-dependent decrease in conductance of an ion channel carrying current outwards, will progressively depolarize the cell membrane towards threshold. This will create a pacemaker potential. The magnitude of depolarization caused by a given decrement in g_K, however, depends upon the total ionic conductance of the membrane (g_T in equation 3.2). The greater the value of g_T the smaller will be the depolarization resulting from a given reduction in g_K. Similarly, a given increment in g_{Na} causes a smaller depolarization the greater the prevailing value of g_T. In resting muscle cells most of the ionic conductance of the plasmalemma is contributed by g_K, so that the depolarizing influence of a given increase in g_{Na} is considerably dependent upon the background level of g_K, as was discussed previously in connection with equation 3.2. The higher the value of g_K the less powerful becomes the depolarizing influence of a given increase in g_{Na}. Note also from Figure 8.1 that the time it takes for a pacemaker potential to reach threshold depends not only upon the pacemaker potential gradient (a), but also upon the value of threshold (b) and the maximal membrane polarity attained (c).

Analysis of conductance changes underlying pacemaking activity, therefore, must include consideration not only of time-dependent conductance changes, but also of the prevailing levels of any background (i.e. non-time-dependent) ion conductances, particularly g_{Na}. As discussed previously in connection with Figures 4.6–4.8, a rise in background g_{Na} will lower the threshold. Figure 8.1b shows that a lower threshold will raise the frequency of action potential generation. Local anaesthetic drugs, by reducing g_{Na}, may thus reduce the frequency of action potential generation. This contributes to the ability of such drugs to correct disturbances of cardiac rhythm which are due to abnormally frequent action potential production, as will be discussed in Chapter 9.

Time-dependent increases in the conductance of slow Ca^{2+} channels probably contribute to some extent to pacemaker potentials in the sinoatrial node, although it is well established that total ionic conductance decreases during diastole, mainly because of a time-dependent decline in the predominant ion conductance, namely g_K. Voltage clamp analysis of the changes in g_K in nodal cells have yielded unexpected results (Figure 8.2). In this experiment sinoatrial nodal cells were voltage-clamped at holding potentials corresponding to the zero-current value (approximately $-45\,mV$) and then step-repolarized (continuous line) to a value in the range $-65\,mV$ to $-50\,mV$. This corresponds to a voltage at which pacemaker potentials would normally arise in non-clamped cells. A post-step time-dependent increase in measured inward current occurred (continuous line). On the other hand, when the nodal cell was depolarized for $100\,ms$ to $+25\,mV$, prior to the repolarization step to $-60\,mV$, in order to simulate an action potential, an outward current flowed during the initial depolarization step. As discussed

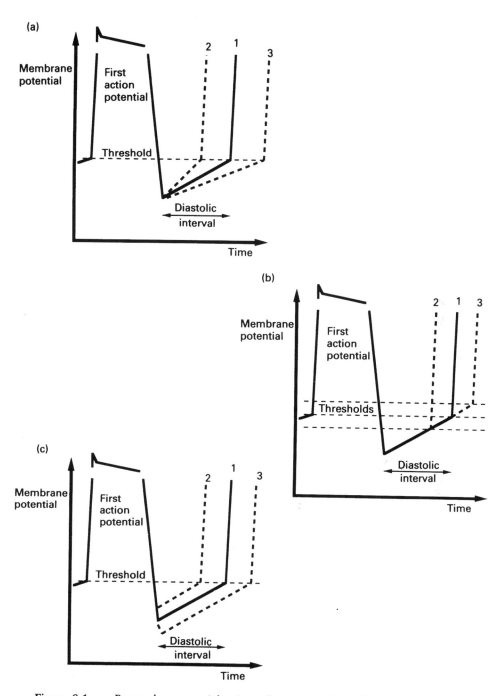

Figure 8.1a–c Pacemaker potentials. A cardiac pacemaker cell showing a first action potential followed by spontaneous diastolic depolarization until membrane potential reaches threshold and a second action potential occurs. Gradients of the diastolic depolarization (8.1a) determine the diastolic interval, as do values of threshold voltage (8.1b) and maximal diastolic membrane polarity reached soon after the first action potential (8.1c).

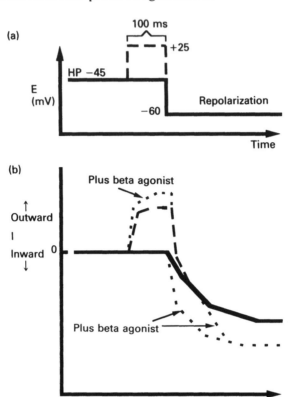

Figure 8.2a–b Pacemaker currents. A sinoatrial nodal cell voltage-clamped, as shown in 8.2a, using a bathing fluid containing blocking drugs for both fast Na^+ and slow Ca^{2+} channels. The cell was clamped at a holding potential (HP) of $-45\,mV$ and then step repolarized to $-60\,mV$ (continuous line in 8.2a), causing inward current (continuous line in 8.2b) to develop time-dependently. A cell depolarized for 100 ms to $+25\,mV$ (dashed lines in 8.2a and 8.2b) displayed outward current. Repolarization from $+25\,mV$ to $-60\,mV$ caused early inward current to develop more rapidly than was seen in the absence of an initial 100 ms depolarization step. Experiments conducted in the presence of a beta adrenoceptor agonist are depicted by dotted lines. This enhanced both outward current (during depolarization) and inward current (during repolarization).

earlier in connection with Figure 4.3, depolarization of a cell usually causes an efflux of K^+, which constitutes an outward current. Note that in the experiment shown in Figure 8.2 the tissue had been treated pharmacologically with a blocker of fast Na^+ channels and a blocker of slow Ca^{2+} channels, so these channels did not contribute. The outward current that flowed during the period of depolarization (dashed line), therefore, seems likely to have been I_K, and the slow build-up of outward current during the depolarization step is consistent with a flow of K^+ through classical delayed outward rectifier channels the outward conductance of which was becoming progressively more activated during this period. Time-dependent offset of this outward I_K would be expected to occur following the subsequent repolarization step to -60 mV. This could account for the 'early' phase of declining outward current (dashed line) that was seen following repolarization. In the absence of a preliminary depolarization step, however, an inward current 'tail' showing a more sluggish time-dependent change occurred following repolarization (continuous line). If the time-dependent inward current was carried by inward I_K one would have to propose that E_K was positive to -60 mV, which is improbable at a physiological extracellular fluid $[K^+]$. More probable would be that a small time-dependently decreasing outward I_K was being increasingly successfully overcome by an inward current carried by ions other than K^+. The rate constants for the changes in total current with time following repolarization obtained with and without the prior depolarization step that are depicted in Figure 8.2, however, were found to be significantly different. Brown and DiFrancesco (1980) proposed, therefore, that two quite distinct time-dependent post-repolarization currents were involved, namely, a time-dependent outward I_K and a time-dependent incrementing inward current, now often labelled I_f, which slowly activates and is then maintained by voltages within the pacemaker range. Most workers have naturally assumed that two distinct channels are involved for these two currents.

The channels through which I_f flows inwards show curiously little ion selectivity, and the predominant inward current-carrying ion under physiological conditions has not yet been identified, although Na^+ is a strong possibility. Another possibility is that the same channel can selectively conduct K^+ outwards (when the electro-chemical gradient is appropriate, as during the depolarization step in Figure 8.2) and conduct other cations inwards (perhaps both Na^+ and Ca^{2+}, when the electrochemical gradient is appropriate, as during the repolarization step depicted in Figure 8.2). The time constants for voltage-induced changes in current might be expected to depend upon which ion was carrying the current. If so, this would explain the differing time constants for inward and outward currents found by Brown and DiFrancesco (1980). It would also explain why reversal potentials for I_f are in the region of -20 mV, which is intermediate between E_K and E_{Na}.

8.2 THE INFLUENCE OF AUTONOMIC NERVE STIMULI ON CARDIAC PACEMAKERS

Study of the modulation of pacemaker potentials by neurohumoral agents secreted by the autonomic nerves has proved particularly informative. The sinoatrial node of a mammalian heart is innervated by both branches of the autonomic nervous system. Sympathetic nerve fibres secrete noradrenaline, which acts upon beta-adrenoceptors and steepens the gradient of the pacemaker potential, so increasing the frequency of action potential development. Parasympathetic nerve fibres secrete acetylcholine. This acts via muscarinic receptors to reduce the pacemaker potential gradient, thus reducing the frequency of action potential development. These changes are illustrated in Figure 8.1a.

The mechanisms whereby neurochemical transmitter substances modify ionic conductances in sinoatrial nodal cells have been studied by many investigators, and the subject has been reviewed by Brown (1982). Muscarinic stimulation of nodal cells increases the outward conductance of time-independent (background or immediately inwardly rectifying) K^+ channels (Carmeliet and Mubagwa, 1986a–c). This makes nodal cells less sensitive to the depolarizing influence of time-dependent increases in conductance of those channels carrying current inwards, as discussed previously in connection with equation 3.2. One time-dependent conductance which possibly activates in the voltage range of sinoatrial nodal pacemaker potentials, and which carries inward current, is that provided by slow Ca^{2+} channels. Muscarinic receptor-stimulation reduces g_{Ca} in these channels, particularly when they are already stimulated by beta-adrenoceptor agonists. By reducing and postponing peak Ca^{2+} conductance in these channels the pacemaker potential gradient is reduced and the threshold is raised, as discussed previously in connection with Figures 4.7 and 4.8.

Beta-adrenoceptor stimulation of nodal cells increases slow Ca^{2+} channel conductance. This conductance activates time-dependently within the voltage range of pacemaker potentials, as described above, steepening the pacemaker potential and lowering the threshold. This action of beta-adrenoceptor stimulation on g_{Ca} is mediated by cyclic adenosine monophosphate-dependent phosphorylation of the ion-channel proteins. Phosphorylation of the delayed outward rectifier K^+ channels also occurs, and this increases their conductance (Iijima et al., 1990; Yazawa and Kameyama, 1990). Time- and voltage-dependent diastolic offset of outward g_K occurs in these channels during the sinoatrial nodal pacemaker potential, as discussed previously in connection with Figure 8.2. The greater the outward g_K that is present in systole, the greater will be the outward K^+ conductance that is available for inactivation during diastole, and so the steeper will be the resulting diastolic depolarization gradient. Of equal, or perhaps greater, physiological significance to the circulatory system is shortening of the cardiac action potential. This also is consequent upon enhanced outward conductance of the delayed outward rectifier K^+ channels as a result of beta-adrenoceptor stimulation. Tachycardia may seriously shorten diastolic filling time of the cardiac chambers unless each systole is suitably shortened by an abbreviated action potential. A third action of beta-adrenoceptor stimulation which is relevant is an enhanced inward I_f produced during diastole when cyclic adenosine monophosphate levels are raised, as shown by the dotted lines in Figure 8.2. The enhanced g_K produced in systole and very early diastole will hyperpolarize the membrane in early diastole into a voltage range where I_f is more fully activated than was previously the case, and so the pacemaker gradient steepens. Beta-adrenoceptor-mediated increases in g_{Ca} of the slow Ca^{2+} channels have particularly prominent effects upon those parts of the myocardium which depend mainly upon these channels for the carriage of inward current at the start of an action potential. The atrioventricular node is the best example: the diastolic membrane potential in this tissue is sufficiently depolarized to effectively inactivate the fast Na^+ channels. The amplitude and rate of rise of the upstroke of action potentials are both increased by beta-adrenoceptor agonists in this part of the heart. Because action potential conduction velocity is dependent upon the rate of change of membrane potential during the upstroke of an action potential, as discussed previously in connection with Figure 4.5, the effect of sympathetic nerve stimulation is to accelerate conduction through the atrioventricular node, which otherwise is the most slowly conducting part of the heart. Conduction velocity is also accelerated by a cyclic adenosine monophosphate-mediated reduction in the electrical resistance of connexons of the intercalated discs, as discussed earlier in connection with Figure 4.5b.

The ability of cyclic adenosine monophosphate to promote phosphorylation of plasmalemmal ion channels and to thereby promote increased ionic conductance is antagonized by an extracellular accumulation of adenosine (Kato *et al.*, 1990; McKinley *et al.*, 1990;

Rankin *et al.*, 1990). This may have physiological importance as a natural restrainer of Ca^{2+} entry and K^+ exit during intense sympathetic nerve stimulation of the heart. When a myocyte becomes overstimulated, to the extent that its nutritional needs are no longer being fully met, there is an extracellular accumulation of the breakdown products of adenosine triphosphate, including adenosine itself. This substance will limit the force of muscle contraction by restraining the rise in $[Ca^{2+}]_i$, and thereby conserve the declining stores of intracellular adenosine triphosphate and prevent excessive loading of the cell with Ca^{2+} or excessive K^+ depletion. This phenomenon may be of greater importance in cardiac myocytes that are strongly contractile than in pacemaker cells of the sinoatrial node.

In summary, pacemaker tissues in mammalian hearts show spontaneous progressive diastolic depolarization due to a time-dependent increase in conductance of ion channels that carry current inwards, probably via I_f, I_{Na} and I_{Ca}, and a time-dependent decrease in conductance of channels that carry current outwards, probably via I_K. Early work on this subject was reviewed by Hauswirth and Singh (1979) and by Carmeliet and Vereecke (1979).

8.3 SLOW-WAVE PACEMAKING

Spontaneous rhythmic fluctuations in membrane potential occur in certain non-cardiac tissues. So-called **slow waves**, for example, are prominent in some smooth muscle types (Carl and Sanders, 1989; Cole and Sanders, 1989a and b; Langton *et al.*, 1989). Background g_K is rather low in these cells. At the maximal level of membrane potential attained (which is only about $-40\,mV$), there is a time-dependent increase in g_{Ca}. This depolarizes the membrane, for reasons discussed previously (equation 3.2). As depolarization proceeds, however, the membrane potential is taken into a range where time-dependent activation of outward g_K occurs, via delayed outward rectifier channels. This halts depolarization and starts repolarization. The rise in g_K is facilitated by a rise in $[Ca^{2+}]_i$, which in turn occurs as a result of the preceding rise in g_{Ca}. As the cell repolarizes, of course, the membrane will enter a voltage range where outward g_K inactivates time-dependently. The whole process will then repeat its cycle. Note that this is quite similar to the cardiac pacemaker potential. Repolarization in a slow wave is due to activation of delayed outward rectifier g_K as a result of the membrane potential having depolarized into a range where this rectifier operates to provide high outward g_K. In the case of cardiac pacemakers, repolarization corresponds to the end of the action potential, but with activation of g_K in delayed outward rectifier channels again a major contributor.

– 9

Disorders of cardiac rhythm

Action potentials normally arise in the sinoatrial node of the mammalian heart. This is where pacemaker potentials have the steepest gradient. Action potentials then spread out to reach all other parts of the heart in the course of a single systole. Disturbances of cardiac rhythm may be considered to arise either from alterations to action potential production (i.e. rhythmicity changes of the heart), or from altered conduction of action potentials through the heart. These possibilities will be dealt with in turn.

9.1 SINUS NODE DYSFUNCTION

Alterations to action potential production can arise because of ionic changes within the sinoatrial node. Enhanced rhythmicity of the sinoatrial node is reflected in an increased frequency of action potential production. This leads to so-called sinus tachycardia, since the heart is being driven at enhanced frequency by the sinoatrial node. Most commonly this occurs in response to increased sympathetic nerve stimulation, as discussed previously in connection with Figure 8.1. The reverse situation, namely sinus bradycardia, occurs most commonly in response to parasympathetic nerve stimulation of the heart. Adjustments in the output of blood from the heart to meet changing needs of the body are partly produced in this way. Sometimes, however, the dominant cardiac pacemaker ceases to be located in the sinoatrial node, and this indicates cardiac disease. If the sinoatrial node stops producing action potentials, or does so only infrequently, a subsidiary and usually slower-acting pacemaker, located in the atrioventricular node, for example, takes over the pacemaking role of the heart. This creates a so-called atrioventricular nodal **escape rhythm**. This is protective, because it serves to maintain a ventricular output of blood, albeit at a reduced level. This emphasizes the fact that some disturbances of cardiac rhythm are protective. For other examples see below.

9.2 ECTOPIC PACEMAKING

If a part of the heart beyond the sinoatrial node generates action potentials at a higher frequency than the sinoatrial node itself, the abnormally located, or **ectopic**, pacemaker will take over as the dominant pacemaker. This may occur transiently, or permanently. The higher the frequency of sinoatrial nodal action potential production, the less likely will an ectopic pacemaker be to reach threshold and generate its own action potential before it is invaded by an action potential delivered from the sinoatrial node.

Ectopic sites of pacemaking can arise anywhere in the heart. Atrial, atrioventricular nodal or ventricular sites, for example, may induce a

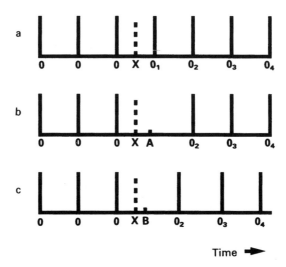

Time ➡

Figure 9.1a–c Compensatory pauses. A regular sequence of ventricular action potentials (O) occurs in an intact heart in response to sinoatrial node-derived action potentials. A single 'extra' electrical stimulus applied to the ventricle (via a pair of wire electrodes) causes a single interpolated ventricular action potential (X) in 9.1a, with subsequent ventricular action potentials (O_1 to O_4) conducted from the sinoatrial node being unaltered in their timing. In 9.1b, a single 'extra' stimulus elicits a ventricular action potential (X) which conducts retrogradely towards the atrioventricular node, where it collides with, and therefore blocks, the next action potential (A) from the sinoatrial node. This causes a so-called **compensatory pause**, after which ventricular action potentials (O_2 to O_4) continue unaltered. In 9.1c, a single 'extra' stimulus elicits a ventricular action potential (X), followed by a shorter compensatory pause than in 9.1b. The retrogradely conducted action potential in 9.1c moved faster than in 9.1b, and reached the sinoatrial node prior to the time (B) when the node would have produced its own next action potential. This resets the nodal pacemaker cycle. Subsequent action potentials (O_2 to O_4) are thus brought forward.

single 'extra' action potential in an otherwise normal train of sinoatrial beats. Alternatively, a few, or in some cases many, such ectopic beats may appear in regular succession. If a single action potential is derived ectopically, say from a site within the Purkinje fibres of the ventricles, the rhythm of the heart can be disturbed in one of three ways, shown in Figure 9.1. In Figure 9.1a, the ectopically-derived premature action potential (X) appears as a simple addition to the train of normal action potentials that were derived from the sinoatrial node (O). Such an extra ventricular systole is said to be interpolated among the normal beats, and ventricular rhythm is minimally disturbed. In Figure 9.1b, on the other hand, action potential X is followed by an abnormally long diastolic pause, known as a **compensatory pause**. This arises because the ectopic action potential (X) conducts retrogradely from the ventricle towards the atria and collides with the next action potential from the sinoatrial node (A). Collision probably occurs within the region of the atrioventricular node. The action potential from the sinoatrial node would fail to reach the ventricles because of this collision, creating a ventricular pause lasting until the next action potential from the sinoatrial node was delivered. The ventricular rhythm here is disturbed not only by the early extra action potential (X), but also by the loss of the next normal action potential (A). Thereafter, however, ventricular rhythm continues as if the extra action potential had not occurred. In Figure 9.1c, the compensatory pause is less prolonged than in Figure 9.1b. Each subsequent ventricular action potential, however, occurs a little earlier than in Figure 9.1b. This is because the retrogradely conducted action potential (X) invaded the sinoatrial node at time B, and this is slightly ahead of the time (A) when the next action potential normally would have arisen in the sinoatrial node. This represents a 're-setting' of the sinoatrial nodal pacemaking cycle by the ectopic action potential.

An uninterrupted succession of ectopically-derived action potentials sometimes arises, usually at a higher frequency than that at which the sinoatrial node normally operates. Two pacemakers can produce action potentials at different rates within the same heart only if there is a block of conduction of those from the faster pacemaker, so that they do not invade the slower pacemaker. The words used to describe different forms of ectopic tachycardia denote their particular ectopic origins (e.g. atrial, atrioventricular nodal or ventricular tachycardia). From a haemodynamic standpoint, the resulting ventricular frequency is the most important factor. The higher the frequency, as a rule, the more compromised the circulation of blood becomes. This is mainly because at high heart rates insufficient time for maximal diastolic filling is provided, so that stroke volume declines. The particular ectopic site of origin is important mainly from a therapeutic standpoint. Drugs which suppress rhythmicity (see below) often display very different potency in different parts of the heart, so that the appropriate choice of suppressive drug depends upon correct identification of the ectopic site of origin. Textbooks of electrocardiography and pharmacology should be consulted for information on this point. As a general rule, when the evidence is ambiguous, it is safer to assume a ventricular origin than a supraventricular origin.

Ectopic rhythmicity of the type so far described in this section probably depends upon ionic processes similar to those responsible for normal pacemaking activity within the sinoatrial node. These were described in detail in Chapter 8. Sympathetic nerve impulses to the heart, therefore, steepen the gradients of such ectopic pacemaker potentials, whereas parasympathetic nerve impulses have the opposite effect. The administration of a beta-adrenoceptor antagonist drug will help to suppress ectopic rhythmicity wherever sympathetic nerve impulses exert a tonic effect upon such ectopic pacemaker sites.

The ability of an ectopic pacemaker to generate an action potential is governed by the gradient of its pacemaker potential and by the threshold voltage, as discussed in Chapter 8 (Figure 8.1). The threshold voltage is dependent upon the conductance of the plasmalemma to inward current-carrying ion channels, notably g_{Na} and g_{Ca}, as discussed in connection with Figures 4.6–4.8. It follows, therefore, that drugs which block slow Ca^{2+} channels or fast Na^+ channels would be expected to raise threshold voltage, and thereby to suppress ectopic pacemaking activity. By and large this is the case. The former group tend to be particularly effective on atrioventricular nodal pacemaking sites and the latter on atrial or ventricular myocardium.

9.3 AFTER-DEPOLARIZATIONS

A distinct type of altered cardiac rhythmicity is created by after-depolarizations of the type described in Chapter 7. These do not occur in normal myocardium. When a muscle cell is overloaded with Ca^{2+}, however, an electrogenic exchange of three extracellular Na ions for each intracellular Ca ion occurs, so that a net inward current flows. This will be maximal during the period of release of stored

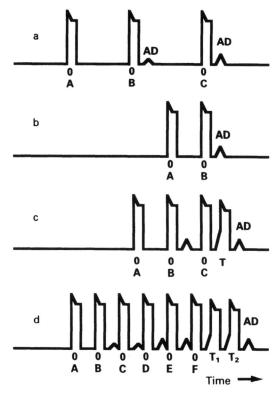

Figure 9.2a–d So-called **after depolarizations** (AD) occurring during a train of electrically stimulated myocardial action potentials (O). In 9.2a, the first action potential (A) was not followed by an AD; the second (B) was, but it was of smaller amplitude than that following the third action potential (C). In 9.2b the frequency of electrical stimulation was higher than in 9.2a, and the amplitude of the AD following the second action potential (B) was higher in 9.2b than in 9.2a. In 9.2c, a higher frequency of electrical stimulation was used than in 9.2a. Following the third action potential (C) the AD reached threshold, causing a **triggered action potential** (T). This was followed in turn by a subthreshold AD. In 9.2d, stimulation frequency was increased further. The AD caused by action potential F and by the first triggered response (T₁) reached threshold, whereas the AD following the second triggered response (T₂) was subthreshold.

Ca^{2+} into the myoplasm, which occurs at the end of each action potential, causing a secondary phasic depolarization (Figure 9.2a). By definition, an after-depolarization occurs after an action potential, but not during quiescence or in the absence of action potentials. Indeed, the phenomenon rarely occurs following the very first action potential in a train (Figure 9.2a), but becomes more prominent with successive members of a train, or with more frequent action potentials in a train. This can be seen by comparing Figures 9.2a and b, and probably represents systolic loading of the cell with Ca^{2+}, which is to some extent cumulative. Such systolic loading occurs in the myocardium of most species, although not all. An after-depolarization may reach threshold, generating an extra action potential, as in Figure 9.2c. Note that the extra beat is triggered by the preceding normal action potential. The triggered action potential itself may induce yet another after-depolarization (Figure 9.2d). If each after-depolarization reaches threshold, a maintained ectopic tachycardia will be produced. This may be of short or long duration. Reduction of Ca^{2+} overload is the logical way to treat such an arrhythmia. If Ca^{2+} overload is due to poisoning with a drug, such as digoxin, withdrawal of the drug will help. Overloading with Ca^{2+} due to ischaemia or stretching of the myocardium may be more difficult to control, but is a worthy therapeutic goal. Drugs which selectively inhibit Na/Ca exchange constitute an attractive possibility for therapy in this situation, although adequate selectivity remains elusive at present.

When the sarcoplasmic reticulum of a myocyte becomes over-loaded with Ca^{2+}, spontaneous release of some, although not all, of the stored Ca^{2+} into the myoplasm may occur. The resulting rise in $[Ca^{2+}]_i$ may be confined to a part of a single myocyte, or it may spread throughout the cell, and even invade adjacent cells. The raised $[Ca^{2+}]_i$ will cause partial activation and shortening of the contractile proteins, unrelated to any action potentials. By means of electrogenic Na/Ca exchange, regions of partial depolarization of the plasmalemma also will occur. This is analogous to the formation of after-depolarizations, except that the former are unrelated to action potentials, and will occur in an unsynchronized manner in different myocytes. Just as after-depolarizations may reach threshold, so also may spontaneous depolarizations, but the latter bear no constant time relationship to preceding action potentials.

9.4 ATRIOVENTRICULAR NODAL DYSFUNCTION

Action potentials conduct through the heart at varying rates, with the slowest velocity normally within the atrioventricular node. This is mainly due to the slower rate of initial depolarization of the action potentials in this part of the heart, compared to other parts, as discussed in connection with equation 4.16. Since atrioventricular nodal action potentials conduct slowly, even under normal conditions, it is not surprising that rhythm disturbances due to conduction impairment are more common in this part of the heart than elsewhere. Figure 9.3 illustrates three degrees of severity of nodal conduction impairment, known as **first, second** and **third degree block**, respectively. In Figure 9.3a there is atrioventricular conduction with a normal delay of **X**, but in Figure 9.3b there is first degree block. In Figure 9.3b each atrial action potential (**A**) conducts, and causes a ventricular action potential (**V**), but the conduction delay (**Y**) is greater than in Figure 9.3a. In Figure 9.3c there is slow conduction of the first atrial action potential (A_1), even slower conduction of the second (A_2), and the third (A_3) fails to conduct, so that $Y < Z < \infty$. Further impairment of conduction by the passage of a preceding action potential quite often occurs, but it is not always displayed as prominently as here. It is probably due to a rise in $[Ca^{2+}]_i$. This type of second degree block creates a 3:2 atrioventricular conduction ratio. Higher ratios produce more profound haemodynamic consequences. In Figure 9.3d there is third degree block; none of the atrial action potentials (**A**) conduct to the ventricles. Ventricular action potentials (**V**) that occur here are due to a ventricular escape pacemaker. The frequency of the escape pacemaker determines the haemodynamic consequences of this nodal disturbance. Too slow an escape pacemaker will seriously decrease cardiac output and hence arterial blood pressure. Note that in Figure 9.3d ventricular action potentials occur regularly, in contrast to the situation in Figure 9.3c. This is an important point of distinction between second and third degree block.

The physiological function of the atrioventricular node appears to be twofold. In the first place, slow electrical conduction through the node introduces a delay between atrial systole and ventricular systole. Atrial systolic ejection of blood occurs prior to ventricular excitation, therefore, and normally contributes modestly to the latter stages of ventricular filling. A second, and probably more important, function of the atrioventricular node is to act as an electrical filter between the atria and ventricles. This arises because of the prolonged refractory period of the node, compared to that of the rest of the heart. This function of the node is not operative during normal rhythm. During atrial tachycardia, however, action potentials may be delivered to the atrioventricular node at a very high frequency. Some arrive during refractory periods created by preceding action potentials, and so all of them do not conduct. Under these circumstances the node displays second degree block, as illustrated in Figure 9.3e. In this particular case alternate atrial action potentials (**A**) conduct. The 2:1 block keeps ventricular systolic frequency

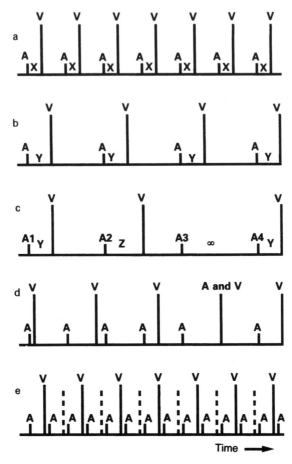

Figure 9.3a−e Atrioventricular nodal conduction. Conduction of a train of action potentials from atria (**A**) to ventricles (**V**) across the atrioventricular node, with a normal conduction time (**X**) in 9.3a, but with a prolongation (**Y**) in 9.3b. This is termed **first degree** atrioventricular nodal block. In 9.3c, the first atrial action potential (A_1) conducted slowly, the second (A_2) conducted even more slowly, and the third (A_3) failed to conduct, so that $Y < Z < \infty$. This is spoken of as **second degree** atrioventricular nodal block, with a $3:2$ atrioventricular ratio. Note that the fourth atrial action potential in 9.3c (A_4) conducted faster than A_2 or A_3. Failure of A_3 to conduct seemed to provide a respite that temporarily improved nodal conduction. In 9.3d, none of the atrial action potentials conducted to the ventricles, and this is spoken of as **third degree** atrioventricular block. The regularly spaced ventricular action potentials in 9.3d were derived from a ventricular **escape pacemaker**. In 9.3e, half of the abnormally frequent atrial action potentials were blocked in the atrioventricular node. This represents an appropriate (physiological) second degree nodal block.

within the physiological range. Had ventricular frequency doubled due to 1:1 conduction through the node, dangerously little ventricular filling time would have been available. Stroke volume may have declined to the point where ventricular output of blood was insufficient to meet the needs of vital organs, notably the brain. The degree of nodal block in Figure 9.3e is both physiologically appropriate and protective to the circulation during an episode of atrial tachycardia. If nodal refractory periods become excessively prolonged, however, ventricular frequency might be insufficient to maintain normal arterial blood pressure, as mentioned in connection with Figure 9.3d.

Atrioventricular nodal action potentials will conduct more swiftly and with less chance of blocking, when the node is being stimulated by sympathetic nervous impulses. This is due to a beta-adrenoceptor-mediated elevation of the levels of cyclic adenosine monophosphate in the nodal cells, causing phosphorylation of three distinct types of ion-conductive pathways, the connexons, the slow Ca^{2+} channels and the delayed outward rectifying K^+ channels. This has three consequences. Firstly, connexons of the intercalated discs acquire enhanced conductance, thereby increasing the space constant, and hence the velocity of action potential conduction through the node, as discussed previously in Chapter 4, particularly in relation to Figure 4.5. Secondly, the plasmalemmal slow Ca^{2+} channels display a faster rising and a higher maximal g_{Ca} value during the upstroke of nodal action potentials. As was discussed previously in connection with equation 4.16, faster rising action potentials will conduct at higher velocity. Thirdly, the delayed outwardly rectifying K^+ channels acquire faster conductance-onset kinetics and reach higher maximal g_K values during the latter stages of an action potential. Shortening of the action potential and its associated refractory periods will occur, as discussed previously in Chapter 7.

Beta-adrenoceptor antagonist drugs will exert actions opposite to those of the sympathetic nervous system listed above. Sometimes, although not always, this may be therapeutically desirable. In a similar way, stimulation of muscarinic receptors in the atrioventricular node usually reduces conduction velocity and prolongs the refractory period. This is due to a reduction in the intracellular concentration of cyclic adenosine monophosphate. Pharmacological blockade of muscarinic receptors in the atrioventricular node, therefore, will cause effects *in vivo* which resemble those of sympathetic nerve stimulation. Where maximal therapeutic relief from pathological atrioventricular nodal block is desired, one may combine the administration of a beta-adrenoceptor agonist (e.g. isoprenaline) with a muscarinic receptor antagonist (e.g. atropine).

9.5 BUNDLE OF HIS BLOCKS

Action potentials sometimes fail to conduct through the Purkinje fibre system of the ventricles in injured or poisoned hearts. The two largest collections of Purkinje fibres are known as the main left and right branches of the bundle of His. These branches are located in the interventricular septum of a mammalian heart. Normally, the delivery of action potentials to left and right ventricles is almost synchronous. The bundle of His branches conduct action potentials at equally high velocity from the atrioventricular node to muscle in the walls of both ventricles. If one main branch is electrically blocked, however, systolic events in the two ventricles begin asynchronously. The ventricle corresponding to a blocked branch of the bundle of His will receive its action potentials late, and from the opposite chamber, by the relatively slow process of cell-to-cell conduction through the ventricular free wall. Electrical events representing total ventricular systole then become abnormally protracted, and this is usually evident in electrocardiographic records from the surface of the body. Treatment is neither necessary nor available.

9.6 ALTERATIONS OF MYOCARDIAL REFRACTORY PERIOD

Myocardial refractory periods undergo biphasic changes during periods of ischaemia. Initially there is a shortening, mainly due to shortening of action potential duration. This latter occurs as a result of declining stores of adenosine triphosphate and a rising $[Ca^{2+}]_i$, as discussed in Chapters 6 and 7. If ischaemia persists for more than a few minutes, however, refractory periods increase again and may considerably outlast the repolarization phase of their action potentials. This is spoken of as **post-repolarization refractoriness**. Action potentials usually remain abbreviated throughout ischaemia. Eventually, refractory periods may exceed those which prevailed before ischaemia began. This often characterizes those areas of chronically ischaemic myocardium that are still electrically excitable. Ultimately, of course, ischaemic tissues may lose their excitability, or even die. Areas of myocardium exhibiting periods of acute ischaemia, co-existent with, but spatially separated from, regions of mild chronic ischaemia, therefore create the greatest unevenness in refractory periods and provide the most arrhythmogenic substrate.

9.7 CONDUCTED RE-ENTRY

Partial conduction failure in a network of muscle fibres can lead to a phenomenon called **re-entry**. This is responsible for many serious disturbances of cardiac rhythm. A simple form of re-entry is depicted in Figure 9.4. A small, perhaps chronically ischaemic region $(X_1 - X_2)$ is envisaged to possess a longer refractory period than that of the remainder of the (perhaps normal) network. If an action potential enters the network at A, soon after a preceding action potential, the tissue at X_1 may still be refractory, so that conduction is blocked. Conduction in the remainder of the network still occurs because of its shorter refractory period. By the time the original action potential arrives at X_2 further time will have elapsed, so the tissue at X_2 may now be re-excitable, enabling a re-entry circuit to be completed. A single action potential may then circulate perpetually in this way.

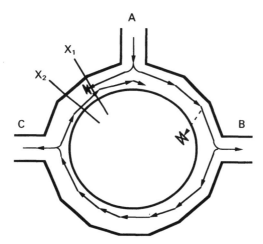

Figure 9.4 An action potential arriving at A conducts towards B, as shown by the arrows, but finds tissue at X_1 still within a refractory period left by a preceding action potential, so it is blocked and extinguished. The action potential that pursued a clockwise route past B and C, however, may reach X_2 after sufficient further time has elapsed for the previously refractory tissue (between X_1 and X_2) to have become re-excitable. The delayed action potential therefore may conduct (retrogradely) through the $X_1 - X_2$ region, completing a circuit. An action potential may continue to circulate or re-enter the path in this way, sending an action potential into the three branches (A, B and C) on each circuit.

Each time it revolves it will re-excite the three branches of the net, thus creating an arrhythmia. In clinical practice, using only body surface electrocardiographic records, it may be difficult or impossible to discriminate between a re-entry tachycardia and one generated by enhanced ectopic rhythmicity. The distinction can be made using mapping studies with electrodes in the heart.

Circuitous re-entry of the type depicted in Figure 9.4 depends upon a critically timed (usually early) action potential. It also depends upon the conduction velocity and refractory period of tissues in the re-entrant pathway. For successful re-entry to occur, an action potential must arrive at A during the period between the end of a shorter and the end of a longer refractory period created by a preceding action potential. The greater the disparity, therefore, between these two refractory periods, the longer the time period or window available for unidirectional block to occur.

Assuming that the refractory periods in both injured and normal parts of the ciruit shown in Figure 9.4 remain unchanged, and that the available re-entry circuit has an anatomically defined or fixed length, the success of re-entry of a unidirectionally blocked action potential will depend only upon conduction velocity. The slower the conduction velocity, the more likely is successful re-entry to occur, and vice versa. Successful re-entry requires that the already unidirectionally blocked action potential conducts retrogradely through the previously blocked area (X_1-X_2). The later that retrograde conduction at X_2 is attempted, the more likely will it be successful, because the more likely will the refractory period at X_2 have ended when the action potential arrives. Although slow conduction increases the chances of successful re-entry, it also reduces the frequency of any resulting re-entry tachycardia. The slower the frequency of a tachycardia, as a rule, the less compromized is the circulation of blood. Hence, slow electrical conduction has both advantageous and disadvantageous consequences. Nevertheless, it is clearly better to have no re-entry than a tolerated re-entry tachycardia.

If conduction velocity remains unchanged, but the refractory periods of injured and normal regions of the network in Figure 9.4 increase by the same amount, the width of the time window available for unidirectional block is preserved. Then, assuming the re-entry circuit is anatomically defined and of fixed length, the chances of a unidirectionally blocked impulse completing the re-entry circuit will diminish. This is because the unidirectionally blocked impulse is now more likely to arrive at X_2 within the extended refractory period set by a preceding action potential.

Summarizing, circuitous re-entry is made more difficult by prolongation of refractory periods and made easier by slowing of conduction velocity. Drugs which prolong refractory periods without slowing conduction velocity, therefore, tend to protect against circuitous re-entry, and vice versa.

Drugs which selectively prolong the refractory period, by blocking Na^+ channels in the way illustrated in Figure 6.2, for example, would be expected to protect against atrial and ventricular re-entry of the type shown in Figure 9.4. At a concentration just sufficient to prolong refractory periods, such a drug would be expected to exert no effect upon conduction velocity. The same drug, however, at higher concentrations, probably would exert actions of the type illustrated in Figures 5.3 and 5.4, and conduction velocity would be impaired. Nevertheless, even at the higher drug concentration, prolongation of refractory periods would be expected to be proportionately greater than other Na^+ channel blocking effects of the drug, so protection against re-entry would persist.

Some Na^+ channel blocking drugs exert actions that are dependent upon the myocyte membrane potential, as shown in Figure 5.4. In general, such drugs are more potent in depolarized tissue. Moreover, most injured myocytes are partially depolarized. The ability of such drugs to block Na^+ channels, therefore, will be greater in injured than in healthy myocardium. Ideally, a drug concentration which is totally inactive on healthy myocardium, but which renders injured myocardium totally electrically inexcitable, may be found. Re-entry would be less likely to occur under these circumstances because it would be confined to tissues displaying more homogeneous electrical behaviour than previously was the case. It is as though the injured area had been electrically eliminated. If the drug prolonged refractory periods in the injured region (X_1-X_2) in Figure 9.4 more than in normal regions, the time window for unidirectional block of an action potential from A would widen. At first sight this seems to be disadvantageous, but note that any unidirectionally blocked action potential also would be more likely to arrive at X_2 within the extended refractory period created by a preceding action potential, so re-entry would actually still fail. In general, therefore, drugs which selectively prolong refractory periods in depolarized tissue are protective against re-entry arrhythmias of the type indicated in Figure 9.4.

9.8 LEADING CIRCLE RE-ENTRY

In Figure 9.4 the central region may represent a hole, such as the orifice of one of the great vessels of the heart, or a refractory patch in a sheet of muscle. In the latter case, an action potential entering at A and circulating around the perimeter will start to spread through the muscle sheet, but may fail to pass right across the central region because the muscle is continuously being rendered refractory by that action potential. This phenomenon is illustrated in Figure 9.4, where the dotted line is blocked on entering the central region, and is spoken of as **leading circle re-entry**. If an occasional action potential does manage to cross the central region, the re-entry path might then become a figure-of-eight, or a double loop. Eventually, the situation might degenerate into multiple loops of ever smaller size and more complex geometry. The minimum length (L) of a successful re-entry loop will be determined by the average conduction velocity (θ) in the loop and by the refractory period (RP) of the most sluggish part of the loop. For re-entry to occur it is necessary that:

$$L > \theta RP \tag{9.1}$$

The cycle time or length of any resulting re-entry tachycardia in a circuit of length L will be L/θ and its frequency will be θ/L.

So far in this account of re-entry circuits the refractory period has been treated as ending abruptly. In reality the absolute refractory period is followed by a partially refractory period (Chapter 6), and the implications of this fact will now be examined. Figure 9.5 depicts an action potential (AP) pursuing a clockwise circuit, with a wavefront at X. The absolute refractory period extends backwards (i.e. anti-clockwise) from X to Y. Tissue in the region YZ is partially refractory. Behind this region is a fully excitable region, usually called the **excitable gap** (EG), the length of which may be expressed in units of distance, or as a time taken by the action potential to traverse that distance. If a drug prolongs the refractory period of tissues in this circuit the length of EG will shorten, as point Z moves anti-clockwise. A point might be reached at which X started to overtake Z and encroach upon partially refractory tissue in the tail of the receding action potential. Action potentials would then arise in still partially refractory tissue. The rate of membrane potential rise in the initial part of the action potential would consequently diminish and action potentials would begin to conduct more slowly, for reasons discussed in connection with equation 4.16. Drugs which selectively prolong refractory periods, therefore, will reduce the

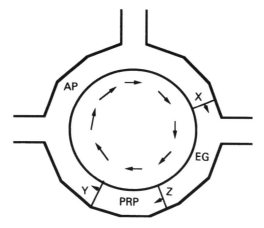

Figure 9.5 Conducted re-entry. An action potential (AP), with a wavefront at X, is circulating around a re-entry pathway in the direction of the arrows. This would arise in the way indicated in Figure 9.4. An absolute refractory period extends backwards (anticlockwise) from X to Y, with a partially refractory period (PRP) from Y to Z, leaving an excitable region, (or gap, EG), where full electrical excitability prevails.

frequency of a re-entry tachycardia before they interrupt it. Slowing of conduction due to selective prolongation of refractoriness is not arrhythmogenic; rather it is a sign that an arrhythmia is being controlled.

The existence of an excitable gap, as in Figure 9.5, is sometimes useful for diagnosis. Clinically, it is often difficult to distinguish between tachycardias arising from enhanced ectopic rhythmicity and those due to re-entry. Electrical stimulation of tissue in the excitable gap of a re-entry circuit with an appropriately timed (early) pulse of current, however, should terminate a tachycardia that is due to re-entry. It would be more likely to fail if the arrhythmia was due to enhanced rhythmicity. Simple electrical stimulation studies of this type are easier to perform than mapping studies, which represent an alternative approach.

9.9 PHARMACOLOGY OF CONDUCTED RE-ENTRY

It follows from equation 9.1 that drug therapy which is designed to interrupt or prevent a re-entry circuit tachycardia should prolong RP (if possible, without reducing θ) to the point where the length of the largest available re-entry pathway (L_{max}) is exceeded by the value of θRP. This will be easier to achieve in small hearts, as in those of the rat, than in those of larger species, such as man. The more minimum re-entry path length is increased, however, the less frequent will any resulting re-entry tachycardia become. This limited goal is worthwhile therapeutically because the slower the tachycardia, as a rule, the less compromised will be stroke volume and thus cardiac output.

Drug therapy should also aim to create uniform electrical properties in previously heterogeneous myocardium. This can operate either temporally or spatially. Most Na^+ channel blocking drugs show a greater potency in tissues that are generating action potentials than in those which are quiescent. This phenomenon, called **use-dependent block**, is more prominent at higher than at lower frequency of stimulation (Figure 6.2). This property may permit such a drug to selectively suppress frequent action potentials arising from a re-entry circuit or from an ectopic region of enhanced rhythmicity, while leaving unaltered much less frequent action potentials, say, of sinoatrial nodal origin. The ability of some Na^+ channel blocking drugs to display a greater potency in depolarized tissue than in normal tissue (Figure 5.4) enables some degree of spatial homogeneity to be achieved in the excitable myocardium. Depolarized, but previously excitable regions may then become totally inexcitable. The remaining electrically excitable tissue will have more uniform properties than before. Note, however, that this homogeneity is achieved at the expense of rendering part of the heart non-excitable, and thus non-contractile. Unfortunately, the resulting weakening of the force of systolic contraction may be more damaging than the original arrhythmia.

Although the existence of zones of myocardium with differing intrinsic electrical properties is particularly arrhythmogenic, as discussed above, theoretically an opportunity for re-entry remains even in structurally entirely homogeneous tissue. Appropriately timed early action potentials may still conduct slowly and then cause re-entry because of regions of variable excitability in the tail of each passing action potential (Davidenko *et al.*, 1990). Perhaps this accounts for some cases of spontaneous re-entry that have been found in hearts which subsequently fail to reveal any structural inhomogeneity.

9.10 FIBRILLATION

Fibrillation can occur in either atria or ventricles. It probably represents the ultimate in degeneration of re-entry loops. Atrial fibrillation involves both atria, just as ventricular fibrillation involves both ventricles. The atrioventricular septum usually prevents spread of fibrillation from atria to ventricles, and vice versa. An exception to this occurs where more than one strand of electrically excitable tissue crosses the atrioventricular septum. Normally, of course, the only electrical connection across the septum is via the bundle of His. A minority of people are born with a subsidiary additional connection.

During fibrillation, action potentials take an ever-changing route through the myocardium. Mapping studies during fibrillation usually reveal several distinct action potential wavefronts occurring simultaneously. Their paths frequently intersect, as each wavefront seeks out and invades still excitable regions of myocardium ahead of itself. If no excitable tissue is found that wavefront is extinguished, but there are usually others which persist. Ventricular fibrillation is often self-limiting in small hearts, but not in man, whereas temporary atrial fibrillation may occur in both small and large hearts. A cardiac chamber which is fibrillating does not eject blood because at any given moment different parts of the wall are at different stages of the systolic-diastolic cycle. In the case of the ventricles, the resulting failure of systolic ejection of blood precipitates immediate total failure of the circulation. Because of the limited time that this state of affairs can be tolerated, particularly by the brain, the only effective cure for this condition is an electric shock, delivered, if necessary, via the chest wall. After a brief asystolic pause normal rhythm is usually resumed. The defibrillatory current shock momentarily depolarizes the entire muscle of the heart. One hopes that when the normal pacemaker tissue resumes its activity the fibrillation will not recur. Atrial fibrillation is far less serious than ventricular fibrillation since atria function mainly as passive conduits, taking blood during ventricular diastole from the great veins to the ventricles. Under normal conditions, atrial systolic activity contributes less than 20% to the total forward propulsive effort of the heart. Indeed, the problem most commonly created by fibrillating atria is the irregular and frequent passage of action potentials via the atrioventricular node and bundle of His to the ventricles. Pharmacological prolongation of the atrioventricular nodal refractory period is usually required to maintain a physiologically normal ventricular frequency in such people.

9.11 REFLECTED RE-ENTRY

So far, re-entry phenomena have been discussed in connection with an action potential conducting around a closed loop of muscle fibres. Re-entry may also occur by reflection across an electrically inexcitable gap (IEG) in a linear cardiac muscle fibre. Figure 9.6 depicts a fibre that has been locally depolarized by injury. The

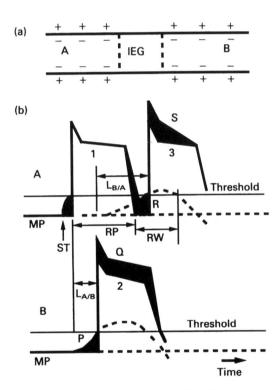

Figure 9.6a–b Reflected re-entry. A myocardial fibre (9.6a) is electrically excitable on either side (A and B) of an inexcitable gap (IEG). Action potential 1 was caused by an electrical stimulus (ST in 9.6b). Dark areas P, Q, R and S represent electrotonic depolarization caused by an action potential on the opposing side of the IEG. Depolarizations P and Q (at B) are due to action potential 1, whereas R and S (at A) are due to action potential 2. The lower borders of the dark areas Q and S represent the membrane potentials (MP) that would have occurred without electrotonic depolarization. Note that these are the same as that of action potential 1. Depolarization Q is superimposed on action potential 2, and S is superimposed on action potential 3. The **reflection window** (RW) represents a time period within which reflection may occur, or the time during which R must reach threshold if reflected action potential 3 is to occur. The refractory period (RP) must end before a RW can begin. Electrotonic depolarization P reaches threshold at the end of latency $L_{A/B}$, and depolarization R reaches threshold at the end of latency $L_{B/A}$.

depolarized central region can no longer generate action potentials, but can still conduct current axially, and hence transmit electrotonic potentials. Action potential 1 in segment A will induce electrotonic depolarization P in segment B. The latter may reach threshold, causing action potential 2 in segment B. Electrotonic latency ($L_{A/B}$) will delay the start of action potential 2, compared with action potential 1. In turn, however, action potential 2 may electrotonically depolarize the tissue at A. If this depolarization (R) reaches threshold another action potential (3) will occur at A. Reflection requires that:

$$L_{A/B} + L_{B/A} > RP \tag{9.2}$$

where $L_{B/A}$ is the electrotonic latency in the direction B to A and RP is the refractory period.

Moreover, Figure 9.6 shows that for reflection to occur there must be a period of time called the **reflection window** (RW), the duration of which is given by:

$$RW > L_{A/B} + L_{B/A} - RP > 0 \tag{9.3}$$

Boundary conditions thus exist for relative values of RP and the sum of $L_{A/B}$ and $L_{B/A}$ if reflection is to occur. This is because electrotonic depolarizations P and R are phasic. That is to say, they rise to a peak and then decline with time as the plateau regions of action potentials 1 and 2 give way to repolarizations. During the latter, outward electrotonic currents flow as shown in Figure 9.6.

Drugs which block myocardial fast Na^+ channels, slow Ca^{2+} channels and the various K^+ channels display a mixture of protective (antiarrhythmic) and deleterious (arrhythmogenic) effects upon reflective processes (Lamanna *et al.*, 1982; Shen and Antzelevitch, 1986). This complexity arises because of the boundary conditions imposed by equation 9.3. Administering a drug which selectively prolongs RP will tend to interrupt an existing reflective re-entry process. During an existing reflective re-entry arrhythmia the value of RP must be less than ($L_{A/B} + L_{B/A}$), as shown by equation 9.2. If ($L_{A/B} + L_{B/A}$) remains unchanged by the drug, a selective drug-induced prolongation of RP may cause the latter to exceed the value of ($L_{A/B} + L_{B/A}$), thereby interrupting the arrhythmia. In practice, of course, many of the drugs which prolong RP by blocking fast Na^+ channels also raise the threshold voltage and thus prolong ($L_{A/B}$ and $L_{B/A}$). It is important, therefore, that latency is not prolonged as much as the refractory period is prolonged if the drug is to succeed in terminating a reflected re-entry arrhythmia. Alternatively, if a drug raises the threshold voltage sufficiently high for electrotonic depolarizations at P or R to not reach threshold, then once again reflected re-entry would be prevented, since latency is now infinite in equations 9.2 and 9.3.

In summary, therefore, so-called antiarrhythmic drugs possess the ability to exert both antiarrhythmic and arrhythmogenic effects, at least under certain circumstances. In general, however, for a patient presenting during an episode of tachycardia that is thought likely to be due to some kind of re-entry process, the administration of a drug with a selective prolonging effect upon refractory periods offers a high probability of terminating the arrhythmic event and restoring sinoatrial rhythm.

9.12 ARRHYTHMIA PROPHYLAXIS

Prophylaxis against the occurrence of arrhythmias is always difficult. When the arrhythmia occurs frequently, but briefly, there is an opportunity to try, in turn, members of a range of different drugs until effective preventative therapy is obtained. If the arrhythmia concerned is rare and self-limiting, it is better not to attempt prophylaxis. If the arrhythmia is not self-limiting, particularly if it is poorly tolerated, or even potentially fatal, it is best to admit the patient to hospital after the first episode and attempt to electrically provoke the arrhythmia with full resuscitation facilities to hand. If reproducibly provocative electrical stimulation parameters can be found, it is then possible to search for an effective prophylactic agent by trial and error. The patient thought to be at risk of developing such an arrhythmia, but who has not yet done so, poses the greatest therapeutic dilemma. Aggressively invasive studies are hard to justify. When techniques of a non-invasive nature become available for studying arrhythmogenesis of the various types discussed in this chapter, we will be in a better position to intervene usefully. The development of such techniques will require investigators to have a thorough understanding of arrhythmogenic processes. This chapter of the book is intended merely as a brief introduction to such processes.

– 10

Epithelial electricity, with special reference to the kidney

Collections of cells arranged as continuous sheets, called epithelia, serve as boundaries between functional fluid compartments, or at the body surface. The primary purpose of epithelia is to regulate water and solute fluxes between compartments, the compositions of which are often different. In some cases, a hydrostatic pressure gradient also exists across epithelia. Since many physiological solutes are ionic in nature, a transepithelial voltage usually exists, for reasons discussed previously in connection with the plasmalemma (equation 1.9).

10.1 EQUIVALENT ELECTRICAL CIRCUIT

Figure 10.1 is the equivalent electrical circuit for a typical epithelium. Three cells are shown, together with their intercellular junctions. Many epithelia are both anatomically and electrically polarized or asymmetric. The side facing outwards is called the **apical** or **mucosal** surface. Reference to an outward surface, however, may be misleading for epithelia that are totally enclosed within the body, and the term mucosal is appropriate only for mucus-secreting epithelia. The most generally applicable word, therefore, is apical. The opposite side, usually adherent to a basement lamina, is variously described as the **inner, serosal** or **basolateral** surface. The last of these terms is the most generally applicable, and implies that basal and lateral walls have properties in common, which they usually do. Specialized junctional structures are commonly present, containing intercellular material, with modifications to adjacent areas of plasmalemma. Junctional complexes serve both to anchor cells together mechanically and to regulate transepithelial electrical and osmotic behaviour.

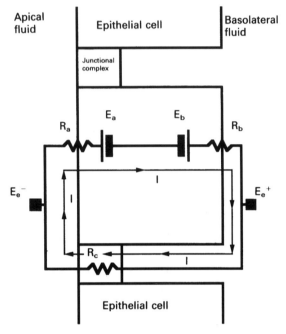

Figure 10.1 Epithelial currents Three epithelial cells are shown, joined at their apical margins by specialized junctional complexes, the electrical resistances of which are R_c. Transcellular current (I) is generated by unequal electrochemical diffusion potentials (E_a and E_b) in apical and basolateral membranes respectively. Current flows transcellularly via R_a and R_b. Overall transepithelial voltage (E_e) is usually apical side negative with respect to the basolateral side.

10.2 TRANSCELLULAR AND PARACELLULAR ROUTES FOR CURRENT

The apical plasmalemma behaves electrically like the membrane depicted in Figure 3.1. Electrochemical diffusion potentials, plus certain other voltages, combine to create an intracellular negativity, measured with respect to extracellular fluid at the apical surface. This is represented in Figure 10.1 by E_a, and is analogous to the membrane potential E in equations 1.9 and 3.2. The apical plasmalemma has a total electrical resistance of R_a. Similarly, the basolateral membrane generates a voltage (E_b) acting via a total resistance of R_b. Because of plasmalemmal asymmetry, however, an overall transepithelial voltage (E_e) occurs across most epithelia, the apical side commonly being negative with respect to the basolateral side. In the thick ascending limb of the loop of Henle, however, this polarity is reversed (below). The electrical circuit in Figure 10.1 is completed by leakage of current across the epithelium via the junctional complex, the resistance of which is R_c.

Applying Ohm's law ($E = IR$) to the transcellular part of the circuit in Figure 10.1, we obtain:

$$(E_b - E_a) - E_e = I(R_a + R_b) \tag{10.1}$$

where I is both the current flowing transcellularly via R_a and R_b, and the current leaking back via R_c.

It follows from Ohm's law that:

$$I = \frac{E_e}{R_c}$$

and this expression for I may be substituted in equation 10.1, giving:

$$(E_b - E_a) - E_e = \frac{E_e}{R_c}(R_a + R_b) \tag{10.2}$$

Rearranging equation 10.2 we obtain:

$$E_b - E_a = \frac{E_e(R_a + R_b)}{R_c} + E_e$$

$$= E_e \left[\frac{R_a + R_b}{R_c} \right] + 1$$

$$= E_e \left[\frac{R_a + R_b}{R_c} + \frac{R_c}{R_c} \right]$$

$$= E_e \left[\frac{R_a + R_b + R_c}{R_c} \right]$$

Hence:

$$E_e = \frac{R_c(E_b - E_a)}{R_a + R_b + R_c} \tag{10.3}$$

10.3 SHORT-CIRCUIT CURRENT

It follows from equation 10.3 that the value of E_e is determined not only by the value of $(E_b - E_a)$, which is a measure of epithelial voltage asymmetry, but also by the value of R_c relative to transcellular resistance $(R_a + R_b)$. The smaller the value of R_c, relative to transcellular resistance, the more **leaky** is the epithelium said to be, and the smaller is the likely value of E_e. In leaky epithelia, the junctional complexes serve to electrically short-circuit epithelial voltage generators. So-called **tight** epithelia, on the other hand, are energetically economical. Although they may generate substantial E_e values, they display low I values unless they are deliberately electrically short-circuited. A short-circuit can be provided, for example, by connecting a zero resistance ammeter between points E_e^+ and E_e^- in Figure 10.1. Short-circuit current (I_{sc}) represents the value of I when E_e is zero in equation 10.1. Hence:

$$I_{sc} = \frac{E_b - E_a}{R_a + R_b} \tag{10.4}$$

Under these conditions, the current flowing through R_c is zero, because there is no transepithelial voltage gradient. Thus I_{sc} flows entirely transcellularly, and is provided by energy-consuming plasmalemmal ion pumps. Steady state I_{sc} is a measure of net pumped ionic flux. In most cells the major pump is the Na/K pump that was described in Chapter 2. In certain specialized locations, however, an H pump and a Ca pump also contribute to the flux.

When an epithelium is not short-circuited, the current flowing transcellularly is matched by an exactly equal transepithelial current in the opposite direction. In Figure 10.1 this countercurrent flows paracellularly via R_c and may be carried by any of the available extracellular ions to which R_c is conductive. This usually means Na^+ and/or Cl^-. Other ions move, but in smaller amounts. Paracellular ionic current is usually accompanied by an osmotically equivalent flow of water, which, in the confined spaces of a junctional complex, may exert a **solvent drag** force upon other extracellular solutes, of both ionic and non-ionic nature. This drag may even move extracellular solutes against an electrochemical gradient, providing another form of secondary active transport.

10.4 PUMPS AND LEAKS

In the epithelium shown in Figure 10.2 a Na^+ leak into the cell occurs on the apical surface, but Na^+ is pumped out of the cell via a Na pump in the basolateral membrane. Transepithelial Na^+ flux is active in the sense that it consumes energy by means of hydrolysis of adenosine triphosphate when operating against a Na^+ electrochemical gradient. Indeed, in many epithelia, in addition to a Na^+ concentration gradient, there is also a voltage gradient to be negotiated, with the basolateral side usually positive in voltage with

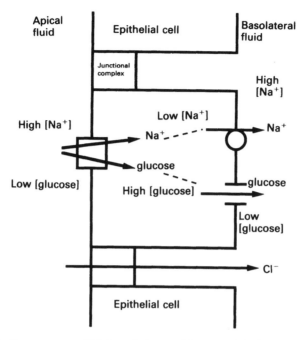

Figure 10.2 Co-transport of Na$^+$ and glucose. Three epithelial cells are shown. An apical plasmalemmal co-transporter of Na$^+$ and glucose is depicted as a small square transected by two arrows. This allows Na$^+$ to diffuse down its electrochemical energy gradient, and at the same time to cause the movement of glucose in the same direction. This raises intracellular glucose concentration to a higher value than in either extracellular fluid compartment, so glucose can passively diffuse out of the cell via the basolateral membrane. A channel through which glucose passes passively (down an energy gradient) is depicted by a gap traversed by an arrow in the basolateral membrane. The Na pump in the basolateral membrane, shown here by the circular arrow, provides the energy not only for Na$^+$ flux across the epithelium, but also for the transepithelial movement of glucose. Paracellular flux of Cl$^-$ constitutes the current (I) flowing via R_c in Figure 10.1. Ultimately it too is driven by the basolateral Na pump.

respect to the apical side. Transcellular net Na$^+$ flux requires another transepithelial ion flux to preserve equality of numbers of anions and cations in the extracellular fluid and to provide a counter-current (I) to complete the kind of circuit shown in Figure 10.1. Flux of Cl$^-$ from apical to basolateral surface, via R_c, serves both of these requirements. Movement of a negatively charged Cl$^-$ is electrically equivalent to flow of positive charge in the opposite direction. In effect, therefore, the epithelium in Figure 10.2 produces net NaCl transport from left to right, while at the same time conforming to the general electrical pattern shown in Figure 10.1.

10.5 CO-TRANSPORTERS

Transepithelial movement of several physiologically important hexose sugars occurs across the walls of kidney nephrons and small intestinal mucosae in the way shown in Figure 10.2. Passive entry of Na^+ via the apical plasmalemma is coupled to flux of sugar in the same direction. Coupling, or co-transport, is necessary for either solute to move, and a lack of either solute will prevent the movement of both of them. Flux of these two solutes across the apical plasmalemma is driven by the existing $[Na^+]$ gradient, which is high outside and low inside. The value of $[Na^+]_i$ is kept below that of $[Na^+]_o$ because of the operation of a basolateral membrane Na pump. Diffusion of Na^+ across the apical plasmalemma makes use of a specialized carrier protein molecule which needs to combine with one Na^+ and one glucose molecule before it diffuses to the inside of the membrane. When the doubly loaded carrier molecule reaches the inner surface of the plasmalemma the low $[Na^+]_i$ that exists in the cytoplasm will allow unbinding of Na^+ from the carrier. This also provokes intracellular unbinding of glucose from the carrier, even when the cytoplasmic glucose concentration is somewhat higher than in the fluid outside the cell at the apical face. In effect, energy has been transferred from Na^+ to glucose molecules via the carrier. Indeed, glucose may be moved into the cell in this way faster than it is metabolized, so the intracellular glucose concentration may rise to exceed that in extracellular fluid at either face of the epithelium. Where this occurs a concentration gradient outwardly directed towards the basolateral surface favours onward diffusion of glucose to the basolateral extracellular fluid. If the concentration of glucose in the basolateral extracellular fluid rises above the intracellular concentration of glucose then the transepithelial flux of glucose ceases, and glucose merely accumulates intracellularly. Co-transport of dietary Na^+ and of glucose from digested food is necessary for absorption of both substances in the small intestine. This is an important consideration therapeutically when using oral rehydration solutions. Maximum absorption depends upon the combined presence of glucose and salt in the bowel lumen.

10.6 CHLORIDE FLUX ACROSS EPITHELIA

Numerous other substances are co-transported by mammalian epithelia. Unlike sugars, most of the other substances are ionic in nature, e.g. amino acids, keto acids, lactate, citrate, phosphate and chloride (Figure 10.3a). This means that transepithelial energy gradients for these substances depend, among other things, upon voltage gradients. Anion movement from apical to basolateral sides is favoured by the usually positive voltage of extracellular fluid on the basolateral side relative to the apical side. Flux of Cl^- is of particular significance because of its abundance in mammalian extracellular fluids, and several ways have evolved for its transport (below). In most cases movement of Cl^- is from apical to basolateral sides, in what is termed an **absorptive** process (Figure 10.3a), although in special cases Cl^- is **secreted** from the apical surface (Figure 10.3b). The major difference here is in the location of the Na/Cl co-transporter, which is apical in Figure 10.3a and basolateral in Figure 10.3b. Figure 10.3a represents a variation on the electrical circuit shown in Figure 10.1. In Figure 10.3a both transepithelial currents pass transcellularly, instead of one of them taking the paracellular route, as in Figure 10.1. In other electrical respects, however, Figure 10.3a is analogous to Figure 10.1. Figure 10.3b is also analogous to Figure 10.1, since electrical current, carried by Cl^-, is passing transcellularly, and Na^+ is carrying counter-current via the paracellular route. In both Figures 10.3a and 10.3b all the K^+ that is being pumped into the cell escapes via the same (basolateral) membrane by which it enters. Many cells, however, have apical membranes containing K^+ channels with a substantial g_K value. Indeed, it is often only the greater transmembrane potential across the apical membrane, compared with the basolateral membrane, that prevents considerable escape of K^+ by the former route. Figure 10.3c depicts an extreme situation, the entire counter-current necessary to balance outward transcellular Cl^- flux being carried transcellularly by K^+. This would constitute a purely KCl-secreting epithelium. In most cases, a mixture of KCl and NaCl secretion occurs by a hybrid of Figures 10.3b and 10.3c. The balance between Na and K fluxes will be determined by the relative g_K values in apical and basolateral membranes, and by their magnitudes relative to g_{Na} in the junctional complexes.

A number of experimentally testable features appear in Figure 10.3a. Inhibition of Na/Cl co-transport, for whatever reason, would be expected to lower $[Cl^-]_i$, and hence to hyperpolarize the cell with

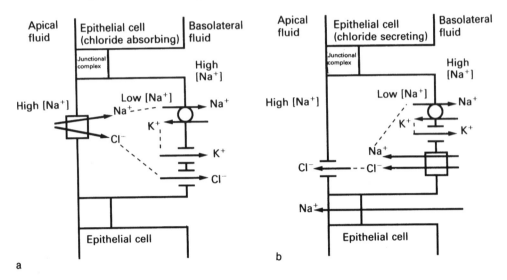

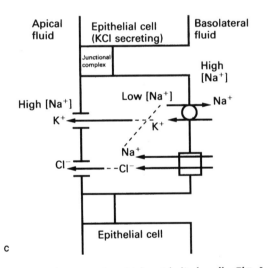

Figure 10.3a–c Contrasting ways in which epithelia handle Cl⁻. In 10.3a, there is net Cl⁻ 'absorption' from the apical surface, whereas in 10.3b there is net Cl⁻ 'secretion' from this face. In both 10.3a and 10.3b the primary energy source for Cl⁻ flux is the Na/K pump in the basolateral membrane. A major difference between 10.3a and 10.3b is the location of the Na/Cl co-transporter. In the Cl⁻ 'absorbing' epithelium (10.3a) it is in the apical plasmalemma, whereas in the Cl⁻ 'secreting' epithelium (10.3b) it is in the basolateral plasmalemma. A major difference between 10.3b and 10.3c is the nature of the ion that provides a counter-current to balance transepithelial Cl⁻ flux. In 10.3b it is Na⁺, moving via the junctional complexes, whereas in 10.3c it is K⁺, moving transcellularly. In 10.3b, the epithelium 'secretes' NaCl, whereas in 10.3c it 'secretes' KCl. From an electrical standpoint 10.3b is exactly the same as Figure 10.1. In 10.3c, however, the current (I, in Figure 10.1) takes a counter-clockwise route around the circuit.

respect to extracellular fluid on either face, as discussed previously in connection with equation 1.9. Because this will increase E_a and E_b equally (equation 10.3) the value of E_e should not change. The easiest way to achieve a reduction of Na/Cl co-transport is usually to reduce extracellular [NaCl]. This, however, tends to reduce I_{sc}, since the Na/K pump becomes less active as $[Na^+]_i$ declines in response to a lower $[Na^+]_o$. The Na/K pump also affects transepithelial Cl^- flux. Thus, an inhibitor of this pump, such as digoxin, will reduce the Cl^- flux by raising $[Na^+]_i$. One of the most important experimental tests of Figure 10.3a, however, is to reduce apical $[Na^+]_o$ without a change in apical $[Cl^-]_o$, or vice versa. Either change would be expected to reduce transepithelial flux of both Na^+ and Cl^- equally.

10.7 THIAZIDE DIURETIC DRUGS

Diuretic drugs of the thiazide class selectively inhibit a Na/Cl co-transporter in apical membranes of some cells lining the tubules of mammalian nephrons. Failure to reabsorb NaCl from the glomerular filtrate creates an enhanced urinary loss of NaCl, with an accompanying increase in the volume of water excreted.

10.8 TRIPLE CO-TRANSPORT OF NA/K/CL

Another way in which Cl^- can move across cell membranes is in association with both Na^+ and K^+, as part of a triple co-transportation process. This was first described in thick segments of the ascending limb of loops of Henle in mammalian kidneys, but it also occurs in various other tissues (Aickin and Brading, 1990a and b). This electrically neutral co-transporter moves one Na, one K and two Cl ions inwards across the apical plasmalemma, as shown in Figure 10.4. In terms of carrier mechanisms, electrical consequences and energetic causes, this form of co-transport is very similar to that described above in connection with Na/Cl co-transport. Diuretic drugs, such as frusemide and bumetanide, inhibit this Na/K/Cl co-transporter. Because this particular form of co-transport is mainly responsible for urinary concentration and dilution in mammals, the administration of such drugs has very profound effects upon water and electrolyte metabolism.

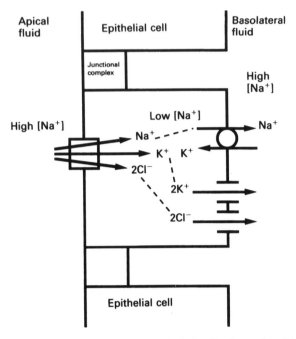

Figure 10.4 Triple co-transport. Three epithelial cells from the thick ascending segment of the loop of Henle are shown. A co-transporter of one Na⁺, one K⁺ and two Cl⁻ is represented by a small square traversed by three arrows in the apical plasmalemma. The net effect created by the apical membrane co-transporter and the Na/K pump in the basolateral membrane is a net transepithelial flux of Na⁺, K⁺ and Cl⁻, from apical to basolateral sides. Any apical loss of K⁺ from the cell would have to be matched by a paracellular Na⁺ movement from the apical side. This is usually slight as few K⁺ channels occur in the apical plasmalemma of these cells.

10.9 ANION EXCHANGERS

In some epithelia, and certain other types of tissue, Cl^- moves across cell membranes in a one-for-one exchange with certain other anions, notably bicarbonate. This was first recognized in red blood corpuscles, where it is well known to be involved in the movement of carbonic acid around the body. Several synthetic drugs inhibit this exchange (Aickin and Brading, 1990a and b). Interestingly, thiazide-type diuretics inhibit this exchange, albeit rather weakly (Ferriola *et al.*, 1986), but it is uncertain to what extent, if any, this contributes to their clinically useful diuretic effect.

10.10 AMILORIDE-INHIBITED NA CHANNELS

Many epithelial cells lining the tubules of mammalian nephrons contain Na^+ channels in their apical membranes, as discussed above. In some of these cells, however, instead of paracellular Cl^- flux, as shown in Figure 10.2, or transcellular Cl^- flux, as shown in Figure 10.3a, it is a transcellular K^+ flux (Figure 10.5) that maintains equality of anions and cations in each compartment (equation 1.2). Inward apical Na^+ flux occurs because of a gradient of $[Na^+]$, which is higher outside than inside, and a voltage gradient, which is outside positive with respect to inside. Once again, the primary source of

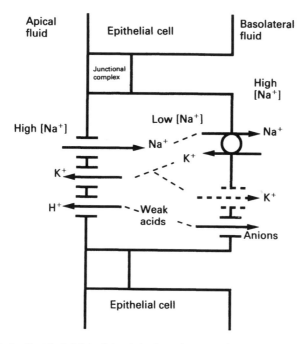

Figure 10.5 Amiloride-inhibited Na^+ leaks. Three epithelial cells from the distal tubule of a mammalian nephron are shown. Na^+ channels in the apical plasmalemma permit influx of Na^+ to occur, unaccompanied by anions. Hence, this Na^+ entry route is depicted as an ordinary ion channel (a gap in the membrane, traversed by an arrow). This contrasts with the apical co-transporter shown in Figure 10.2. A basolateral membrane Na/K pump ejects Na^+ to the extracellular fluid, and actively transports K^+ inwards. Subsequently, K^+ passively diffuses out of the cell, down its electrochemical energy gradient. The magnitude (and direction) of K^+ energy gradients on the two faces of the cell are determined, among other things, by transmembrane potentials on these two faces.

energy for ionic movement is the basolateral Na/K pump, which keeps $[Na^+]_i$ low. The basic electrical difference between Figures 10.2 and 10.5, therefore, lies in the route taken by circuit-completing electric current of the type depicted in Figure 10.1. In Figure 10.2 transcellular current (carried by Na^+) completes an electric circuit by means of an equal flow of Cl^- current via paracellular junctional complexes (R_c). In Figure 10.5, on the other hand, it is transcellular K^+ flux that provides the right to left counter-current (I) needed to complete the circuit shown in Figure 10.1. But what determines which species of ion provides the counter-current? As discussed above in connection with Figure 10.3c, the route of escape of K^+ from an epithelial cell is determined by two factors, namely, by g_K in the two cellular faces and by the electrochemical energy gradients for this ion across the two faces. If a high conductance and a favourable energy gradient exist for K^+ movement in a direction which provides for a right to left current in Figure 10.5, which is to say, for exit of K^+ from the apical surface, then this ion and this route will provide the counter-current needed to balance the left to right transcellular flux of Na^+. On the other hand, in so far as K^+ has to leave the cell mainly via basolateral membranes because this is energetically favoured, the counter-current needed for Figure 10.1 would be a flow of some other ion. It could be a left to right paracellular Cl^- flux, for example, as in Figure 10.2. In the case of Figure 10.5, however, a high g_{Na} in the apical plasmalemma will electrically depolarize this side of the cell, making exit of K^+ easier from this side (equation 3.2). Moreover, apical membrane g_K is present. Hence K^+ becomes a counter-current-carrying ion, and the epithelial cell becomes K^+ secretory. In effect, the apical membrane acts here as an indirect K/Na exchanger. Diuretic drugs of the amiloride type inhibit apical membrane g_{Na} in certain renal tubular cells. The resulting inhibition of Na^+ reabsorption and K^+ secretion make these drugs particularly useful clinically in the manipulation of electrolyte balance, especially when a patient's store of K^+ has been or is in danger of becoming depleted.

10.11 PROTON SECRETION

The situation discussed so far in connection with Figure 10.5 is somewhat over-simplified, however, since most cells have other intracellular cations besides K^+ with which to generate a right to left counter-current across the apical plasmalemma. The most important such alternative source of charge is the proton (H^+). Distal tubular epithelial cells of the nephron contain the enzyme carbonic anhydrase, which catalyses a reaction between carbon dioxide and water, producing carbonic acid (H_2CO_3). Ionization of the latter substance provides an intracellular source of H^+, as indicated in Figure 10.5. Note, however, that this only serves as a source of right to left counter-current, of the type shown in Figure 10.1, if HCO_3^- leaves the cell via the basolateral membrane, but this usually occurs because of preferential apical membrane depolarization caused by a high g_{Na}.

The presence of carbonic anhydrase in the apical membrane of renal tubular cells has the effect of converting HCO_3^- in the glomerular filtrate, where it occurs at a concentration of about 25 mEquiv/l, into carbon dioxide and water. When the carbon dioxide produced has diffused into the cells, which it does readily, it is enzymatically reconverted to H_2CO_3. Indeed, this constitutes a major route for HCO_3^- reabsorption from glomerular filtrate. In view of the central place of HCO_3^- in acid-base balance in the blood plasma, this renal process assumes great physiological significance.

In the light of the above facts it is hardly surprising that in mammalian nephrons apical influx of some Na^+ is balanced, in the electrical sense, by both H^+ and K^+ counter-movement, with the former predominating. Nevertheless, secretion of K^+ is necessary from a physiological standpoint, as it provides the major hormonally regulated process for controlling K^+ loss into the urine. This is needed in order to exactly balance intestinal absorption of K^+ from the diet. The adrenal gland hormone, aldosterone, increases both the number and the conductance of apical plasmalemmal Na^+ channels. This hormone, therefore, not only promotes Na, and particularly $NaHCO_3$ reabsorption from the glomerular filtrate, but also promotes K^+ and H^+ secretion into the urine.

10.12 CATION EXCHANGE CARRIERS

Exchange of various other cations, notably Na/H and Na/Ca, across cell membranes occurs in many mammalian tissues, including renal tubules. The former has been investigated by Khandoudi *et al.* (1990). The latter was discussed previously in Chapters 7 and 9. These are all passive co-transport processes that are driven by a favourable (or 'downhill') energy gradient for at least one of the participating ions. An ion moving 'down' such an energy gradient can then transfer energy, via the carrier protein molecule, to another ion. Thus the latter is enabled to move 'up' an energy gradient, in the way that was discussed previously in this Chapter in connection with Na/glucose and Na/Cl co-transport. The main difference is that during a cation/cation exchange the two types of cation move in opposite directions. The energy gradients concerned in the exchange are usually electrochemical, i.e. contain both solute concentration and voltage elements. Being reversible, as a rule, these co-transporters may function in either direction. In the case of the Na/Ca exchanger, for example, this means that it is possible for Ca to enter cells in exchange for Na, or for Na to enter in exchange for Ca. A rise in $[Na^+]_i$, or a fall in $[Ca^{2+}]_i$, will promote both Ca influx and Na efflux, and vice versa, as discussed in Chapters 7 and 9.

Thus, manipulation of the relative magnitudes and directions of the transmembrane concentration gradients for the two cations can determine the prevailing direction and rate of ionic exchange. Transmembrane voltage also has an effect upon the Na/Ca co-transporter. This is because the carrier protein moves a different number of positive charges in opposing directions across the membrane per unit of time. As pointed out in Chapter 9, three monovalent Na ions move across the membrane for each divalent Ca ion that moves. A single net positive charge, therefore, moves with the carrier in an opposite direction to that taken by each Ca ion. The carrier protein to which the cations bind seems to be able to move across the membrane only when it is fully loaded with ions, i.e. with three Na^+ on one side and one Ca^{2+} on the other side. Moreover, when fully loaded, the burden of positive charges will be asymmetric, with a higher positive charge on the Na-loaded side. The positively charged side of the carrier will be attracted to, and thus move faster towards, the side of the membrane at the more negative voltage. At the same time, of course, the Ca-loaded side is made to move at a faster rate towards the side of the membrane that is at the more positive voltage. Transmembrane voltage might be expected to retard or inhibit movement of the fully loaded carrier, therefore, when Na is moving towards the side of the membrane that is at the more positive voltage. This also means that, under the same circumstances, the exchanger is inhibited when Ca is moving towards the side of the membrane that is at the more negative voltage. Most mammalian cells at rest have an interior that is negative with respect to the cell exterior, and this will tend to oppose Na efflux. Depolarization of such a cell, therefore, would be expected to promote Na efflux, and hence Ca influx. This has been shown to occur in many types of mammalian cells (Kimura *et al.*, 1987; Beuckelmann and Wier, 1989; Li and Kimura, 1990).

In summary, electrical activity in epithelia has been described in this chapter in terms which are closely analogous to those used for the plasmalemma in Chapter 3. The similarity is most easily seen by comparing Figures 3.1 and 10.1. In both cases electric current flows continuously around a circuit, due to the presence of electrochemical diffusion potentials and ionic exchange currents. Different ions carry the current at different points in the circuit, but the current has the same magnitude at all points in the circuit. Just as it is meaningless to try to study, or describe, the inward current in Figure 3.1 in isolation from the associated outward current, so it is important to consider the right to left paracellular counter-current in Figure 10.1 at the same time as studying the transcellular fluxes of ions. The latter tend to dominate our attention, probably because they represent physiologically regulated and pharmacologically manipulatable processes. Any alteration to the latter, however, must also produce an alteration to the former.

The early literature on epithelial electricity has been comprehensively reviewed by Giebisch and Windhager (1973) for renal tubules, and by Curran and Schultz (1968) and Schultz and Curran (1968) for gastrointestinal mucosae.

References

Aickin, C.C. and Brading, A.F. (1990a) Effect of Na^+ and K^+ on Cl^- distribution in guinea-pig vas deferens smooth muscle: evidence for Na^+, K^+, Cl^- co-transport. *J. Physiol.*, **421**, 13–32.

Aickin, C.C. and Brading, A.F. (1990b) The effect of loop diuretics on Cl^- transport in smooth muscle of the guinea-pig vas deferens and taenia from the caecum. *J. Physiol.*, **421**, 33–53.

Attwell, D., Cohen, I., Eisner, D., Ohba, M. and Ojeda, C. (1979) The steady state TTX-sensitive ('window') sodium current in cardiac Purkinje fibers. *Pflugers Arch.*, **379**, 137–42.

Berger, F., Borchard, U. and Hafner, D. (1989) Effects of (+)- and (±)-sotolol on repolarizing outward currents and pacemaker current in sheep cardiac Purkinje fibres. *Naun. Schmied. Arch. Pharmacol.*, **340**, 696–704.

Beuckelmann, D.J. and Wier, W.G. (1989) Sodium-calcium exchange in guinea-pig cardiac cells: exchange current and changes in intracellular Ca^{2+}. *J. Physiol.*, **414**, 499–520.

Brown, H.F. (1982) Electrophysiology of the sinoatrial node. *Physiol. Rev.*, **62**, 505–30.

Brown, H.F. and DiFrancesco, D. (1980) Voltage clamp investigations of membrane currents underlying pacemaker activity in rabbit sinoatrial node. *J. Physiol.*, **308**, 331–51.

Campbell, D.L. and Giles, W. (1990) Calcium currents, in *Calcium and the Heart*, (ed. G.A. Langer), Raven Press, New York, pp. 27–83.

Carl, A. and Sanders, K.M. (1989) Ca^{2+}-activated K channels of canine colonic myocytes. *Am. J. Physiol.*, **257**, C470–80.

Carmeliet, E. and Mubagwa, K. (1986a) Changes by acetylcholine of membrane currents in rabbit cardiac Purkinje fibres. *J. Physiol.*, **371**, 201–17.

Carmeliet, E. and Mubagwa, K. (1986b) Characterization of the acetylcholine-induced current in rabbit cardiac Purkinje fibres. *J. Physiol.*, **371**, 219–37.

Carmeliet, E. and Mubagwa, K. (1986c) Desensitization of the acetylcholine-induced increase of potassium conductance in rabbit cardiac Purkinje fibres. *J. Physiol.*, **371**, 239–55.

Carmeliet, E. and Vereecke, J. (1979) Electrogenesis of the action potential and automaticity, in *Handbook of Physiology*, Sect. 2, Vol. I, (ed. R.M. Berne) American Physiological Society, Bethesda, Maryland, pp. 269–334.

Clark, R.B., Giles, W.R. and Imaizumi, Y. (1988) Properties of the transient outward current in rabbit atrial cells. *J. Physiol.*, **405**, 147–68.

Colatsky, T.J., Follmer, C.H. and Starmer, C.F. (1990) Channel specificity in antiarrhythmic drug action. Mechanism of potassium channel block and its role in suppressing and aggravating cardiac arrhythmias. *Circulation*, **82**, 2235–42.

Cole, W.C. and Sanders, K.M. (1989a) Characterization of macroscopic outward currents of canine colonic myocytes. *Am. J. Physiol.*, **257**, C461–9.

Cole, W.C. and Sanders, K.M. (1989b) G proteins mediate suppression of Ca^{2+}-activated K^+ current by acetylcholine in smooth muscle cells. *Am. J. Physiol.*, 257, C596–C600.

Curran, P.F. and Schultz, S.G. (1968) Transport across membranes: general principles, in *Handbook of Physiology*, Sect. 6, Vol. III, (ed. C.F. Code), American Physiological Society, Washington, DC, pp. 1217–43.

Davidenko, J.M., Kent, P.F., Chialvo, D.R., Michaels, D.C. and Jaliffe, J. (1990) Sustained vortex-like waves in normal isolated ventricular muscle. *Proc. Nat. Acad. Sci. USA*, 87, 8785–9.

Donnan, F.G. (1924) The theory of membrane equilibria. *Chem. Rev.*, 1, 73–90.

Dukes, I.D. and Morad, M. (1989) Tedisamil inactivates transient outward K^+ current in rat ventricular myocytes. *Am. J. Physiol.*, 257, H1746–9.

Dumaine, R., Schanne, O.F. and Ruiz-Petrich, E. (1990) Taurine depresses I_{Na} and depolarises the membrane but does not affect membrane surface charge in perfused rabbit hearts. *Cardiovasc. Res.*, 24, 918–24.

Ferriola, P.C., Acara, M.A. and Duffey, M.E. (1986) Thiazide diuretics inhibit chloride absorption by rabbit distal colon. *J. Pharmacol. Exp. Therap.*, 238, 912–15.

Fozzard, H.A. (1979) Conduction of the action potential, in *Handbook of Physiology*, Sect. 2, Vol. I, (ed. R.M. Berne), American Physiological Society, Bethesda, Maryland, pp. 335–56.

Gettes, L.S. and Reuter, H. (1974) Slow recovery from inactivation of inward currents in mammalian myocardial fibres. *J. Physiol.*, 240, 703–24.

Giebisch, G. and Windhager, E.E. (1973) Electrolyte transport across renal tubular membranes, in *Handbook of Physiology*, Sect. 8, (eds J. Orloff and R.W. Berliner), American Physiological Society, Washington, DC, pp. 315–76.

Giles, W.R. and Imaizumi, Y. (1988) Comparison of potassium channels in rabbit atrial and ventricular cells. *J. Physiol.*, 405, 123–45.

Grover, J.G., Sleph, P.G. and Dzwonczyk, S. (1990) Pharmacologic profile of cromakalim in the treatment of myocardial ischemia in isolated rat hearts and anesthetized dogs. *J. Cardiovasc. Pharmacol.*, 16, 853–64.

Haas, H.G., Kern, R., Einwachter, H.M. and Tarr, M. (1971) Kinetics of Na activation in frog atria. *Pflugers Arch.*, 323, 141–57.

Harvey, R.D. and Ten Eick, R.E. (1988) Characterization of the inward-rectifying potassium current in cat ventricular myocytes. *J. Gen. Physiol.*, 91, 593–615.

Hauswirth, O. and Singh, B.H. (1979) Ionic mechanisms in heart muscle in relation to the genesis and the pharamcological control of cardiac arrhythmias. *Pharmacol. Rev.*, 30, 5–63.

Hirano, Y. and Hiraoka, M. (1986) Changes in K^+ currents induced by Ba^{++} in guinea pig ventricular cells. *Am. J. Physiol.*, 251, H24–33.

Hiraoka, M. and Kawano, S. (1986) Contribution of the transient outward current to the repolarization of rabbit ventricular cells. *Jap. Heart J.*, 27, Suppl. I, pp. 77–82.

Hiraoka, M. and Kawano, S. (1989) Calcium-sensitive and insensitive transient outward current in rabbit ventricular myocytes. *J. Physiol.*, 410, 187–212.

Hodgkin, A.L. (1954) A note on conduction velocity. *J. Physiol.*, 125, 221–4.

Hodgkin, A.L. and Huxley, A.F. (1952) A quantitative description of membrane current and its application to conduction and excitation in nerve. *J. Physiol.*, 117, 500–44.

Hodgkin, A.L. and Rushton, W.A.H. (1946) The electrical constants of a crustacean nerve fibre. *Proc. Roy. Soc. Series B*, **133**, 444–79.

Hondeghem, L.M. and Katzung, B.G. (1977) Time- and voltage-dependent interactions of antiarrhythmic drugs with cardiac sodium channels. *Biochimica Biophys. Acta*, **472**, 373–98.

Hondeghem, L.M. and Katzung, B.G. (1984) Antiarrhythmic agents: the modulated receptor mechanism of action of sodium and calcium-channel blocking drugs. *Ann. Rev. Pharmacol. Toxicol.*, **24**, 387–423.

Hutter, O.F. and Noble, D. (1960) Rectifying properties of heart muscle. *Nature*, **188**, 495.

Iijima, T., Imagawa, J.-I. and Taira, N. (1990) Differential modulation by beta adrenoceptors of inward calcium and delayed rectifier potassium current in single ventricular cells of guinea-pig heart. *J. Pharmacol. Exp. Therap.*, **254**, 141–6.

Kass, R.S., Arena, J.P. and Walsh, K.B. (1990) Measurement and block of potassium channel currents in the heart. *Drug Development Research*, **19**, 115–27.

Kato, M., Yamaguchi, H. and Ochi, R. (1990) Mechanism of adenosine-induced inhibition of calcium current in guinea-pig ventricular cells. *Circulation Res.*, **67**, 1134–41.

Khandoudi, N., Bernard, M., Cozzone, P. and Feuvray, D. (1990) Intracellular pH and the role of Na^+/H^+ exchange during ischaemia and reperfusion of normal and diabetic rat hearts. *Cardiovasc. Res.*, **24**, 873–8.

Kilborn, M.J. and Fedida, D. (1990) A study of the developmental changes in outward currents of rat ventricular myocytes. *J. Physiol.*, **430**, 37–60.

Kimura, J., Miyamae, S. and Noma, A. (1987) Identification of sodium-calcium exchange current in single ventricular cells of guinea-pig. *J. Physiol.*, **384**, 199–222.

Kurachi, Y. (1986) Ionic mechanisms underlying the negative slope formation of the inward-rectifying potassium background current in ventricular cell of guinea-pig heart. *Jap. Heart J.*, **27**, Suppl. I, 65–72.

Lamanna, V., Antzelevitch, C. and Moe, G.K. (1982) Effects of lidocaine on conduction through depolarized canine false tendons and on a model of reflected re-entry. *J. Pharmacol. Exp. Therap.*, **221**, 353–61.

Langton, P.D., Burke, E.P. and Sanders, K.M. (1989) Participation of Ca currents in colonic electrical activity. *Am. J. Physiol.*, **257**, C451–60.

Li, J. and Kimura, J. (1990) Translocation mechanism of Na – Ca exchange in single cardiac cells of guinea-pig. *J. Gen. Physiol.*, **96**, 777–88.

Lombert, A. and Lazdunski, M. (1984) Characterization, solubilization, affinity labeling and purification of the cardiac Na^+ channel using Tityces toxin Y. *Eur. J. Biochem.*, **141**, 651–60.

McCullough. J.R., Conder, M.L. and Griffel, L.H. (1990) Electrophysiological actions of BRL 34915 in isolated guinea pig ventricular myocytes. *Drug Development Res.*, **19**, 141–51.

McKinley, J.B., Dahlman, D. and Macloed, K.M. (1990) The interaction of adenosine analogues with cAMP-generating and cAMP-independent positive inotropic agents in rabbit left atrium. *Naun. Schmied. Arch. Pharmacol.*, **342**, 605–12.

Naharashi, T. (ed) (1990) *Ion Channels*, Vol. 2, Raven Press, New York.

Nakayama, T. and Fozzard, H.A. (1988) Adrenergic modulation of the transient outward current in isolated canine Purkinje cells. *Circulation Res.*, 62, 162–72.

Noda, M., Shimizu, S., Tanabe *et al.*, (1984) Primary structure of *Electrophorus electricus* sodium channel derived from cDNA sequence. *Nature*, 312, 9121–7.

Noble, D. and Stein, R.B. (1966) The threshold conditions for initiation of action potentials by excitable cells. *J. Physiol.*, 187, 129–62.

Rankin, A.C., Sitsapesan, R. and Kane, K.A. (1990) Antagonism by adenosine and ATP of an isoprenaline-induced background current in guinea-pig ventricular myocytes. *J. Mol. Cell. Cardiol.*, 22, 137–8.

Sakmann, B. and Trube, G. (1984a) Conductance properties of single inwardly rectifying potassium channels in ventricular cells from guinea-pig heart. *J. Physiol.*, 347, 641–57.

Sakmann, B. and Trube, G. (1984b) Voltage dependent inactivation of inward-rectifying single channel currents in the guinea-pig heart cell membrane. *J. Physiol.*, 347, 659–83.

Sanguinetti, M.C. and Jurkiewicz, N.K. (1990) Two components of cardiac delayed rectifier K^+ current. Differential sensitivity to block by Class III antiarrhythmic agents. *J. Gen. Physiol.*, 96, 195–215.

Scanley, B.E., Hanck, D.A., Chay, T. and Fozzard, H.A. (1990) Kinetic analysis of single sodium channels from canine cardiac Purkinje cells. *J. Gen. Physiol.*, 95, 411–37.

Scher, A.M. and Spach, M.S. (1979) Cardiac depolarization and repolarization and the electrocardiogram, in *Handbook of Physiology*, Sect. 2, Vol. I, (ed. R.M. Berne), American Physiological Society, Bethesda, Maryland, pp. 357–92.

Schultz, S.G. and Curran, P.F. (1968) Intestinal absorption of sodium chloride and water, in *Handbook of Physiology*, Sect. 6, Vol. III, (ed. C.F. Code), American Physiological Society, Washington, DC, pp. 1245–75.

Shen, X.S. and Antzelevitch, C. (1986) Mechanisms underlying the antiarrhythmic and arrhythmogenic actions of quinidine in a Purkinje fiber-ischemic gap preparation of reflected re-entry. *Circulation*, 73, 1342–53.

Shibasaki, T. (1987) Conductance and kinetics of delayed rectifier potassium channels in nodal cells of the rabbit heart. *J. Physiol.*, 387, 227–50.

Sommer, J.R. and Johnson, E.A. (1968) Cardiac muscle: a comparative study of Purkinje fibers and ventricular fibers. *J. Cell Biol.*, 36, 497–526.

Sperelakis, N. (1979) Origin of the cardiac resting potential, in *Handbook of Physiology*, Sect. 2, Vol. I, (ed. R.M. Berne), American Physiological Society, Bethesda, Maryland, pp. 187–267.

Tohse, N. (1990) Calcium-sensitive delayed rectifier potassium current in guinea pig ventricular cells. *Am. J. Physiol.*, 258, H1200–7.

Trautwein, W. (1963) Generation and conduction of impulses in the heart as affected by drugs. *Pharmacol. Rev.*, 15, 277–332.

Weidmann, S. (1955) Rectifier properties of Purkinje fibers. *Am. J. Physiol.*, 183, 671.

Wilde, A.A.M., Escande, D., Schumacher, C.A., Thuringer, D., Mestre, M., Fiolet, J.W.T. and Janse, M.J. (1990) Potassium accumulation in the globally ischaemic mammalian heart. A role for the ATP-sensitive potassium channel. *Circulation Res.*, 67, 835–43.

Yazawa, K. and Kameyama, M. (1990) Mechanism of receptor-mediated modulation of the delayed outward potassium current in guinea-pig ventricular myocytes. *J. Physiol.*, **421**, 135–50.

Yue, D.T. and Marban, E. (1988) A novel cardiac potassium channel that is active and conductive at depolarised potentials. *Pflugers Arch.*, **413**, 127–33.

Index